PNEUMATIC
CONVEYING

PNEUMATIC CONVEYING
Second Edition

H. A. STOESS, JR., P.E.

Material Handling Engineer (Retired)
East Allen Township, Northampton, Pennsylvania

Distributed in cooperation with
Powder and Bulk Engineering,
Minneapolis, MN

WILEY

A Wiley-Interscience Publication

JOHN WILEY & SONS

New York · Chichester · Brisbane · Toronto · Singapore

Library of Congress Cataloging in Publication Data:

Stoess, H. A., Jr.
    Pneumatic conveying.

    "A Wiley-Interscience publication."
    Bibliography: p.
    Includes index.
    1. Pneumatic-tube transportation.   2. Bulk solids
handling.   I. Title.
TJ1015.S75   1983   621.8′672   83-6915
ISBN 0-471-86935-X

Printed in the United States of America

10 9 8 7 6 5 4 3 2

PREFACE

Since the first edition of this book was written in 1969, many advances have been made in the application of pneumatic conveyors. Innovations in pipeline design and methods of operation have resulted in reduced pipeline velocities in medium- and high-pressure systems and the conveying of abrasive materials. Proprietary information now abounds within the engineering structure of many suppliers of such equipment. Mention of these systems is made here, however, to indicate their application and availability.

This book was written with five objectives: (1) to present the types of pneumatic conveyors available to handle the many different bulk materials: the first three chapters describe the conveyors, enumerate the various types, and present information to determine the type, design, arrangement, and equipment descriptions; (2) to present research methods, with some of their problems; (3) to illustrate with drawings and photographs, the application of pneumatic conveyors to industries, including descriptive matter on what to do and what not to do; (4) to list and describe bulk materials applicable for handling in pneumatic conveyors, including the characteristics that determine their conveyability and flowability; and (5) additions and deletions to the First Edition to bring this volume as up-to-date as possible.

I am variously indebted to over one hundred companies and individuals who answered my correspondence and telephone inquiries. Their many courtesies and valuable contributions in the form of photographic material, technical data, catalog and pricing information, and material characteristics contributed greatly to the preparation of this second edition. I must especially thank those who offered many words of encouragement and well wishes; to those who advised the First Edition was mandatory reading for new employees; and to those lecturers and educators who advised they used the First Edition as reference and teaching material.

Finally, I must express great personal satisfaction in my writing and typing the manuscript and proofreading the galley and page proofs, all of which I hope will have been well done.

H. A. STOESS, JR.

East Allen Township
Northampton, Pennsylvania
May 1983

CONTENTS

PNEUMATIC CONVEYING

1

WHAT IS A PNEUMATIC CONVEYOR?

1.1 GENERAL DESCRIPTION

Pneumatic conveying may be defined as the art of transporting bulk materials through a pipeline by either a negative or a positive pressure airstream. It is also described as the harnessing of air movement to accomplish work. This airstream or air movement is nothing more than air in motion. When in motion, air becomes wind, and does such things as gently waving growing grain, pushing ships, or even devastating miles of forests.

The simplest pneumatic conveyor is only a miniature hurricane, directed to a certain destination by means of a pipe, with the movement of air caused by introducing it into the pipe at a pressure above normal [14.7 psi (pounds per square inch) at 70°F and sea level] or by lowering the pressure below normal at the delivery end. In either way, this air moves to regain a normal or absolute pressure and, when moving with sufficient velocity, will carry objects in its path, depending on its substance or weight.

Air has a weight of 0.075 lb/cu ft (pounds per cubic foot) at 70°F and an absolute pressure of 14.7 psi. As the pressure is lowered below absolute pressure, following the law of gases, air loses weight as well as viscosity. This reduces the impinging or carrying capacity of material through the pipeline. Conversely, if the pressure in the pipeline is above absolute pressure of 14.7 psi, again following the law of gases, the air is heavier; thus it has greater impinging and carrying capacity, enabling it to convey more material. This is the basis of design within the art.

The reference to art is made here because pneumatic conveying is more an art than a science. Truthfully, it is the trickiest of all of the bulk material

1

handling arts. The introductory paragraph of the 1917 catalog of the Pneumatic Conveyor Company, Chicago, stated:

> This catalog contains cuts showing a few of the different types of pneumatic installations made by us during the past six years, together with descriptive matter covering the essential features and advantages to be found in the pneumatic unloading of cars and vessels and the conveying of materials to remote storage. The simplicity and efficiency of these systems depends entirely on accurate design based on a comprehensive knowledge of pneumatics in general and an intimacy with the characteristics of the material in question, and more valuable than both, actual experience.

The art even today lies in intimacy with the characteristics of the material in question when it is under the influence of air.

Much effort has been spent by the industry in attempting to correlate information that would permit design by formulas. The result, for the most part, has been unsuccessful. This is due to the almost infinite variety of physical characteristics of materials and arrangement of equipment. Many times, materials of the same name and general appearance have a wide variety of conveying charcteristics. For example, some materials will range from approximately 20 to 50 lb/cu ft, and their moisture content will vary from 0 to 1% up to the point where pneumatic conveying becomes impractical. Some other materials will range from 70 to 110 lb/cu ft. The only answer to this problem is experience, with much common sense based on that experience. Add to this the problem of abrasiveness and crystal breakage or degradation. The free silica content, among other characteristics of materials being conveyed, will determine the degree of abrasiveness with the resultant difference in type of equipment required. Fluidity under the influence of air will also have a bearing on the equipment required. A friable material calls for special consideration in the choice of velocity and the denseness of the conveying stream. An improperly designed pneumatic conveyor can be the best grinder or pulverizer on the market, although not the most economical. There is *no* universal pneumatic conveyor made that will handle all materials, hence the need for and application of the types in use today.

1.2 ADVANTAGES

In the design and layout of pneumatic conveyors, the elimination of straight-line conveying is an accomplished fact. Most mechanical conveyors, when change in direction is necessary, require transfer points. This means that a series of conveyors is required, which adds to their cost and to control problems. With a pipeline as the conveying structure, short and long radius bends are utilized to change direction, avoid interferences, and require minimum supporting structures. Similarly minimized are both space requirements and the accesses needed for maintenance purposes.

Housecleaning or cleanliness is improved immeasurably by pipeline conveying. Pipe joints are tight, whether they be welded or flanged joints or pipe lengths connected by compression couplings. Similarly, if a vacuum system is used, the atmospheric pressure inside the pipeline is lower than that outside the pipeline; thus all leakages are in, not out. In pressure systems, tight joints allow for air pressures up to 100 psig inside the pipeline with no loss of dust or material, minimizing housekeeping. The relatively few field joints eliminate dust and dirty conditions. It must be remembered that housekeeping in one's plant is an expense that may be considerable and yet may not be under scrutiny of the cost control expert. Moreover, freedom from dust, plant cleanliness, and low handling loss or shrinkage of the material being conveyed add not only good plant appearance, but also savings in materials through the suppression of dust losses.

Pneumatic conveyors offer greater safety to employees than any other type of bulk conveyor. Many estimates have been made of the cost to American industry of accidents during the handling of bulk materials. Recent published figures indicate that approximately 22% of all compensible injuries occur while materials are being handled. It is estimated that wage loss, medical expense, and insurance costs associated with handling accidents in one year amounted to approximately $350 million and the so-called indirect costs to about $225 million. Thus, for both the "specified" and the "indirect" costs, the annual national total runs about $575 million. It is obvious, therefore, that management, as well as material-handling and industrial engineers, cannot afford to show lack of interest in more efficient and safer handling techniques. There is no doubt that the accident problem can be substantially corrected through a careful analysis of the extent of the problem. From this standpoint, pneumatic conveyors can be considered whenever their application is feasible. Information on the frequency and severity rates is limited; however, one report for a single state for the year 1951 gave pneumatic conveyors and unloaders no fatal accidents and 136 nonfatal accidents. For mechanical conveying as a whole, the report gave 6 fatalities and 648 nonfatal accidents. For mechanical car unloading and loading, it gave 2 fatal accidents and 229 nonfatal accidents. In the construction of a cement plant, before the establishment of a safety program by one of the trade associations, the Insurance Underwriters advised that the fluid solids pumps (Fuller-Kinyon pump) in the plant eliminated the source of 52% of all lost-time accidents that were attributed to screw conveyors, bucket elevators, and their drives.

In addition to creating safer working conditions, pneumatic conveyors have reduced fire and explosion hazards. Among the combustible materials conveyed through pneumatic conveyors are wood flour, starch, flour, cellulose acetate, and gunpowder pellets. In 1947, tests were conducted to determine the fire hazard involved in high-velocity dust–air mixtures of cellulose acetate in a conveying pipe. Under the conditions of the tests, it was concluded that a local combustion would not propagate throughout the length of the system. It was also concluded that conditions of velocity and turbulence must modify combustion characteristics so that maximum explosion pressures and rates are not experienced.

One rating bureau specializing in the feed mill and grain elevator trade extends a sizable credit in insurance rates to feed and flour mills employing pneumatic stock-handling system exclusively. Their daily contact with the mill and elevator trade has made them familiar with the lessened fire hazards of pneumatically equipped mills. To qualify for this rate reduction, all materials-handling systems handling bulk materials must be pneumatic and in noncombustible spouting. In flour mills utilizing pneumatic handling for the milling operation but nonpneumatic means for a cleaning department, and which are not cut off by fire walls from the mill, a smaller rate credit is available. Feed mills must employ all pneumatic conveying to receive a rate reduction.

1.3   DISADVANTAGES

Advantages in the use of pneumatic conveyors are many compared to the disadvantages, which have lessened considerably with each succeeding year. Manufacturers and users are learning more about the art, thereby improving reliability. In years past, one of the strongest criticisms has been the higher power costs compared to other types of conveyors. Much research, however, is aimed at reducing this objection.

In many cases, initial capital investment is higher than that for other types of conveyors. This, however, can be offset by the labor-saving quality and possible reduction in insurance rates. A minor disadvantage is the ability of this conveyor to convey in one direction only (i.e., it is unidirectional).

1.4   LIMITATIONS

The principal limiting factor in the use of pneumatic conveyors is usually the material to be conveyed. Materials to be conveyed, prior to the recently advanced development of the blow tank, had been classed as dry and relatively free flowing. Since those developments, sluggish, abrasive, and damp materials are now being conveyed with little or no difficulty. The free-flowing characteristic should be weighed very carefully, since some materials, although not free-flowing (sluggish) in the common sense of the term, can be made relatively free-flowing through the influence of aeration.

Friable materials should not be conveyed pneumatically, except possibly the low-velocity, dense phase type of pressure (blow tank) system. An exception occurs when partial degradation of the material being conveyed is inconsequential to its end use. Many of the spray-dried products, which are usually in prilled form, cannot be pneumatically conveyed without, at times, excessive degradation. Normally, the prilled form is fragile, and if it is to be maintained, its degree of hardness and structural stability should be investigated.

Conveying distance may limit the use of pneumatic conveyors. A greater development of the art, however, has made vacuum systems practical up to 1800 ft in length, low-pressure systems practical up to and over 1 mi in length, and high pressure practical up to 10,000 ft, all without booster stations.

2 TYPES OF PNEUMATIC CONVEYORS

The term "pneumatic conveyor" is commonly used to describe two methods of transporting bulk materials by means of compressed air power, negative or positive. The first method is the production of a low-, medium-, or high-velocity airstream into a pipeline system, suspending the material in the airstream. The second method is aerating or fluidizing the material and then forcing the fluidized column of material through a pipeline system by the energy of expanding compressed air. The basic equipment operating under these two methods may be divided into nine distinct types partly by reference to their air requirements, which may range from a few cubic feet of air per minute to substantial volumes, at pressures measured in inches of water to as high as 125 psi and vacuums measured from 6 up to 12 in. Hg (inches of mercury).

The nine types of pneumatic conveyors are (1) vacuum, using a high-velocity airstream to suspend the material in a pipeline system at velocities ranging from 4000 to 8000 fpm (feet per minute), and at vacuums up to 12 in. Hg; (2) low-pressure, using a medium-velocity airstream to suspend materials in the pipeline system at pressures up to 12 psig; (3) medium-pressure, using a low-velocity airstream up to 45 psig; (4) high-pressure, again using a low-velocity airstream up to 125 psig; (5) high density, pulse system, using an extremely low-velocity airstream at pressures up to 30 psig; (6) low-pressure, venturi feed system, using medium- to high-velocity airstreams, but a low pressure of 1 or possibly 2 psig; (7) a combination of vacuum and any of the low-, medium-, or high-pressure systems; (8) closed-circuit, using both low to medium velocities and pressures; and (9) air-activated gravity conveyors, which use low-pressure air to aerate or fluidize a pulverized material, at least to a degree sufficient to fluidize the material in immediate contact

7

with a slightly inclined air-permeable surface, to overcome friction with it and with the side walls of the conveying duct, whereby the mass will flow by gravity.

2.1 VACUUM SYSTEM

The vacuum system is used when conveying from a multiplicity of pickup points or points of origin to a single delivery destination. The feeding of material into a pipeline having a negative pressure (i.e., less than 14.7 psia) is a relatively simple matter, whatever the feeding arrangement used. The vacuum system is also used for the unloading of dry bulk material from railroad box and covered hopper cars, as well as the specialty cars, barges, and ships. Materials conveyed include dry pulverized, crushed granular, and, under certain conditions, materials having a particle size as large as 2 in. With today's knowledge, systems up to approximately 1800 ft long have been installed and successfully operated.

The vacuum system, very much like the home vacuum cleaner, utilizes an air intake, a material intake, or a combination of both, a material transport line, an air–material separator, and the power plant. In conveying from a multiplicity of pickup or originating points (Figure 2.1), a metered feed to the system is needed when feeding is done from a head of material. To minimize air leakage into the system the metering device must be relatively airtight, especially when more than one such device is used to feed material

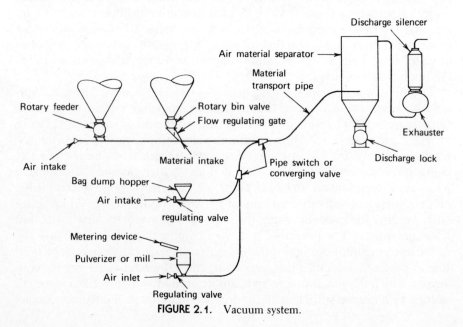

FIGURE 2.1. Vacuum system.

into a single pipeline. If substantial air leakage is allowed, air volumes required will increase to maintain conveying, causing larger equipment than necessary downstream from the point or points of intake.

The most frequently used feeding device is the rotary feeder. This type of feeder has a star wheel rotating in an enclosed housing with closely machined tolerances. Driven by a slow-speed gearmotor through a roller chain drive, the volume of material delivered to the airstream in the pipeline can be closely controlled. It is used on aerated materials capable of flushing, materials having irregular particle size and shape, and where several materials, or different amounts of material, are delivered to the airstream.

When grains or materials having a regular particle size are delivered to the airstream, an open-and-shut bin valve, together with a spout with a flow-regulating gate, is used. The open-and-shut valve merely allows the material either to flow or not to flow. The flow-regulating gate, which is nothing more than a variable orifice, meters the material into the conveying airstream. In handling grains, a rotary feeder used under this condition will tend to clip the kernels, causing some degradation to the grain.

When bags of material are emptied into the conveying system, a bag dump hopper with a grizzly at its top is used. The grizzly, a coarse screen or screen bars, keeps paper and the like from entering the conveying system. Paper and similar items will foul equipment downstream and must be kept out of the conveying system. Usually, no metering device or flow-regulating gate is used since bag dumping is not, by any means, a continuous feed. Its intermittent quality allows the correctly designed system to swallow up the bag of material, digest it in the conveying airstream, and deliver it to its destination with extremely little or no trouble. To minimize the dusty conditions attributed to bag dumping, either part or all the conveying air for the system is drawn into the conveying pipeline through the bag dump hopper. An in-draft of air at the top of the hopper minimizes the amount of dust that will fly around the area.

The vacuum system is well suited for conveying from the discharge of mills or pulverizers, since some or all of the required conveying air can be drawn through the mill or pulverizer, air sweeping the unit to prevent dusting. A metering device should be used ahead of the mill or pulverizer not only to regulate the feed to the conveying system but also to avoid overloading the mill or pulverizer.

The material, or discharge from the feeding device, is mixed with the airstream in the material intake. From this point to the air–material separator, the mixture of air and material moves through the pipeline of straight tangents and long radius bends. At the termination of the pipeline, the air–material mixture enters the air–material separator. At this point, the material is separated from the airstream. The material discharges from the separator through a discharge lock, with the air going to the exhauster of the power plant.

The air–material separator primarily is a cyclone receiver with a conical section sufficient to permit the material to discharge through the discharge lock attached to the material outlet flange at the bottom. When conveying relatively dustless materials, one cyclone is at times sufficient to separate the material from the air. When additional separation is required, a two-stage receiver, one cyclone within the other, or two separate cyclone receivers in series, is used. With cyclone receivers having efficiencies in the 90–95% range, dusts will pass through the receivers and then the exhauster to atmosphere. If slight air pollution together with a dust exhaust can be tolerated, this may be sufficient. Where this condition cannot be tolerated, and complete separation—100% visible retention of dust and material from the exhaust—is required to meet Environmental Protection Agency (EPA) requirements, a combination filter–receiver is used.

The filter-receiver will have a cyclone bottom section and an upper section equipped with a multiplicity of cloth or synthetic filter bags to retain the fine particles. The filter bags are continuously cleaned by either reversal of air through the bags, mechanical shaking, or a combination of both.

At the material outlet of the filter-receiver, a discharge lock is used to discharge the material from the bottom cone; at the same time, this seals the system against loss of conveying air and vacumm. This lock, either a rotary feeder or gate lock, should be sized to discharge the material at a rate at least twice that of the incoming material from the pipeline. Bulking of material due to aeration should also be taken into consideration. Material should not be allowed to build or back up in the cone of the filter-receiver.

For nonabrasive materials, rotary feeders having machined cast iron bodies and close-fitting fabricated cast iron rotors are used. For abrasive materials and materials having a tendency to smear or build up on closely fitted rotating surfaces, the three-compartment gate lock is used.

A motion safety switch should be connected to the tail shaft of either the rotary feeder or gate lock to indicate rotation and operation of the unit. This safety switch is interlocked electrically with the electric motor driving the exhauster, so that in the event the feeder or gate lock is inoperative, material will back up in the air–material separator to the point where it will foul the filter media, break through it, and then continue on through the exhauster to atmosphere. By shutting down the exhauster, the whole system becomes inoperative, thereby keeping material from being discharged to atmosphere and, just as important, protecting the exhauster and its continued operation.

From the discharge at the top of the air–material separator, clean air should be emitted and piped to the inlet of the exhauster. This exhauster is a rotary positive blower used as a vacuum pump, usually driven by an electric motor through a multiple V-belt drive. A suitable silencer, to meet Occupational Safety and Health Administration (OSHA) requirements, should be installed as close as possible to the discharge of the exhauster to minimize the noise level of the machine. From the discharge of the silencer, the air is then relieved to atmosphere.

In car unloading, the vacuum system is well suited to the transfer of material from the car to storage or process points (Figure 2.2). For boxcars, where the material is completely at rest, a flexible hose of sufficient length to reach each end of the car is required. Attached to the end of the hose is a nozzle whose function is to pick up the material at a regulated rate and to mix the material with the conveying air for transport through the pipeline to the air–material separator.

In boxcar unloading, someone must enter the car and unload the material. In this type of operation, the maximum size hose one worker can handle efficiently is 5 in. If hose of larger size is used, the operator finds it more difficult to handle, hence efficiency is reduced with a lower overall unloading or conveying rate the result. In some instances, where a high unloading rate is desired, either two 5-in. hoses or a single 7-in. hose in connection with an 8-in. material transport pipe is used. Although two 5-in. hoses can be used in a boxcar, a 7-in. hose without mechanical assistance is impossible. A telescopic riser, raised and lowered by an electric winch, assists the operator in getting the hose in and out of the car as well as helping him extend the hose the length of the car (Figure 2.3). To further assist the operator, the nozzle of the 7-in. hose is equipped with a dolly with casters. In actual operation, a single 7-in. hose is capable of handling about 15–20% more material than two 5-in. hoses. The two 5-in. hoses, to be efficient, must convey materials simultaneously. If one hose is allowed to run light (no material conveying), it will admit more air to the system, starving the other hose of much-needed air and vacuum.

In covered hopper car unloading, attachments that allow above-rail operation are available, obviating the under-track pit. Single pans that attach directly to the outlet gates of the car connect to a single hose. If two hoses are used, two pans are used. A covered hopper car unloading unit (Figure 2.4) rests atop and across the rails, attaching to the outlet gates through flexible connections. One hose can be used, or two hoses can be used simultaneously with this unit. In operation, when both sides of the car are being unloaded and two hoses are used, one side of the car will invariably be emptied before the other. This, however, is for only a short period of time and does not materially affect the total unloading time.

Vacuum systems should be given careful consideration when cleanliness, freedom from dust, low handling loss or shrinkage, good labor relations, and greater safety and reduced fire and explosion hazards are of great concern. Cleanliness is fostered because material leakage is into, not out of, the point of such leakage. Low handling loss or shrinkage can also be attributed to in-instead of out-leakage. Good labor relations are especially significant in the unloading of materials from railroad cars, where operations relatively free from dust are more the rule than the exception. Greater safety

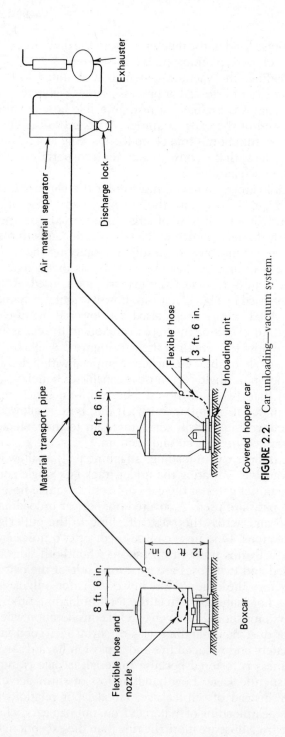

FIGURE 2.2. Car unloading—vacuum system.

Exhauster

Discharge lock

Air material separator

Material transport pipe

Flexible hose

3 ft. 6 in.

Unloading unit

Covered hopper car

8 ft. 6 in.

12 ft. 0 in.

8 ft. 6 in.

Flexible hose and nozzle

Boxcar

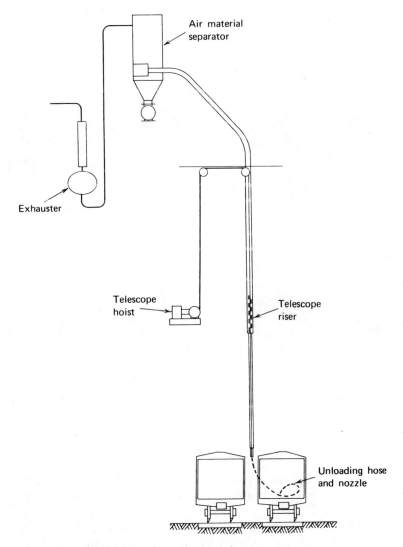

FIGURE 2.3.   Car unloading using telescopic riser.

and reduced fire and explosion hazards can be attributed to the minimum amount of moving machinery involved at the point of operation and the inhibition of the vaccum system to propagate an explosion or explosive mixture.

2.2 LOW-PRESSURE SYSTEM

Low-pressure systems are categorized as such by the limitation imposed on the combination of the air supply and feeding mechanism. Rotary positive

FIGURE 2.4. Twin-nozzle covered hopper car unloading unit. (Courtesy Fuller Company, Bethlehem, Pennsylvania).

blowers of the lobe type are used primarily to activate the system. This type of blower is limited in output pressure to 12 psig. Rotary feeders and gate locks, depending on type, will stand differential pressures across them from a few pounds per square inch to 20 psig. The low-pressure system (Figure 2.5) is used in in-plant or process conveying and car unloading, conveying from a single pickup point or intake point to a multiplicity of discharge points. Materials conveyed include dry pulverized, crushed granular, and fibrous materials.

The greater emphasis being put on dust suppression and reduction in air pollution necessitates special study of this type of system at both the inlet to the system and the discharge, especially when dusty materials are being handled. At the inlet to the system, material at atmospheric conditions (14.7 psia) is passed through the feeding device into the conveying airstream, which is under a positive pressure of up to 26.7 psia. With a rotary feeder as the feeding mechanism, compressed air fills the pockets of the rotor as it discharges the material into the conveying zone. As the rotor rotates and reaches the inlet opening, this compressed air expands upward through the inlet, causing puffs of air to aerate and disturb the material, causing a dust nuisance. Dust suppression equipment must be used to contain the dust within the confines of the equipment. Figure 2.6*a* illustrates equipment requirements for conveying from the outlet of a bin. Figure 2.6*b* suggests

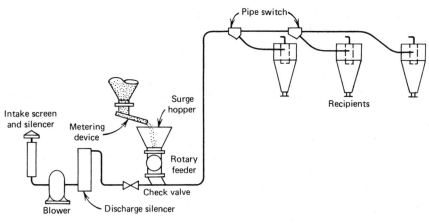

FIGURE 2.5. Low-pressure system.

attaching the rotary feeder directly to the outlet flange of the bin, which would certainly confine the blowback air and accompanying dust, and, it is hoped, would dissipate up through the material in the bin. It would be fine if the material in the bin could absorb this air and dust and still maintain flow out of the bin. Ninety-nine times out of a hundred air dispersion will not happen. What does happen is that the blowback air will not dissipate fast enough up through the material, causing air pressure to build up at the outlet of the bin. The air pressure will build up to the point where it will hold up the material in the bin, preventing it from discharging from the bin into the pockets of the rotary feeder. The result is a no-flow condition and a no-conveying operation.

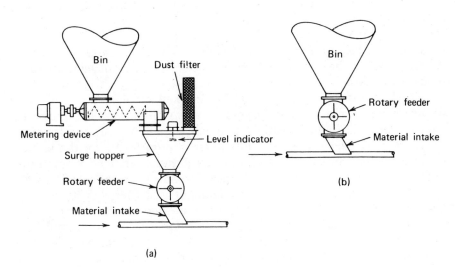

FIGURE 2.6. Feeding material from bin into low-pressure system.

The blower to activate the system is of the lobe type, driven by an electric motor through a multiple V-belt drive. Attached to the inlet of the blower is an intake screen and silencer. The screen keeps large objects from entering the silencer and blower, and the silencer suppresses the amount of noise emitted by the blower. For optimum silencing, a silencer should also be installed at the blower discharge. This silencer, as well as the intake silencer, should be as close as possible to the respective flanges of the blower to achieve maximum results.

If the material being conveyed in the system must be protected against contamination from foreign materials, the intake screen is refined to include a filter media to restrain contaminables from entering the blower and system. Further refinement is to install an in-line filter after the blower discharge, to protect the material being conveyed from contamination from rust or other material that may occur within the cast iron housings and impellers of the rotary blowers and other appurtenances.

Between the outlet of the blower or other appurtenances, piping is used to deliver the air to the material intake of the system. In this piping, a check valve should be installed. The purpose of this check valve is to prevent blowback of material and air into the blower in the event of a power failure or other malfunction. This blowback and malfunction may never occur— it is just good sense and insurance against possible breakdown of the blower unit, one of the two most vital parts of the system. The other vital part is the feeding mechanism, which feeds material into the airstream.

The feeding mechanism is usually some type of rotary feeder (described more fully in Section 3.3). At this point a few words on feeder effectiveness in the low-pressure system are in order. Be sure to use a feeder constructed so that it minimizes the leakage of air. Abnormal or large amounts of air leakage is a costly operation. Whatever air leaks out of the system through the feeder is lost to the conveying operation and the efficiency of the system. The blower must deliver not only the air required for conveying but also the air lost by leakage through the feeder. If leakage is great, power and operating cost for the operation are increased with no benefit accruing from the increased cost.

A good case in point is the system using a small-diameter (2 in. and smaller) pipeline and operating at close to maximum allowable pressure. Such systems will require possibly as much as 160 cu ft air/min for conveying. If the feeder leakage reaches 40 cu ft air/min, 20% of the blower output is wasted. Suppose wear has increased the clearances between the rotor and the body of the feeder. Leakage could then easily double to 80 cu ft air/min. With no increase in the amount of air delivered by the blower, this added leakage reduces conveying air by as much as 25%. This added reduction in air available to the pipeline can cause a no-conveying situation through the reduction in air velocity within the conveying pipeline. The larger the pipeline system, the less it is affected by feeder leakage. In many cases, it is desirable to increase the size of the pipeline system to decrease the operating pressure, thereby reducing feeder leakage. For conveying the same amount of material, while operating pressure will be lower, larger air volume will be required, but the power required by the blower will be about the same.

From the discharge of the feeder, the material is mixed with the conveying air in the material intake tee. The energy of expanding compressed air then propels this mixture through the pipeline to the terminal points. The terminal points may number one or many, either bins, tanks, process vessels, weigh hoppers, or railroad cars.

Rotary feeders are normally used in feeding only nonabrasive to mildly abrasive materials. A design of rotary feeder having high-hardness cast metal rotors and wearing blocks in the upper quadrants and with adjustable-for-wear features does handle abrasive materials. This type of feeder is applicable for both low- and high-pressure systems. It should have, however, a metered feed to limit pocket filling. For abrasive materials, and where the system pressure reaches a maximum of 15 psig, the three-compartment gate lock is applicable. In operation, the three-compartment gate lock system is a continuous type of system. It will handle both pulverized and grandular materials.

In the usc of the gate lock for feeding materials into the airstream, care must be exercised in feeding the material to the gate lock (Figure 2.7) and in discharging it into the pipeline. In feeding to the gate lock, a metering feed is required, since the lock cannot operate under a head of material. If a head of material greater than the volumetric displacement between the gates minus the operating space is allowed to build up above the top gate, the material will overfill the space, preventing the gates from operating.

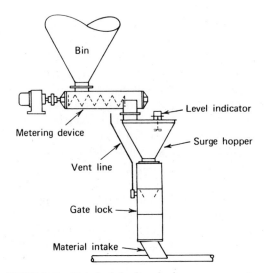

FIGURE 2.7. Gate lock feeding for low-pressure system.

To a certain degree, the gate lock might be termed an intermittent feeder, since it is designed to discharge material approximately nine times per minute. In contrast, the rotary feeder has up to 120 discharges per minute out of its pocketed rotor. This reduction in number of discharges means a higher volume of each discharge when comparable rates are handled, causing turbulence and possible plugging in the air–material intake to the system. Either a conveying pipeline size of sufficient diameter to absorb the higher volume of material or a flow-regulating device between the outlet flange of the gate lock and the air–material intake is required.

Another design employs two chambers in series (similar to the gate lock), but with self-cleaning sliding disk valves for alternately opening and closing the chambers for filling and discharging. This airlock can be used not only for feeding material into a pressure system up to 15 psig, but also as a discharge lock below the air material separator of a vacuum-type system having a maximum vacuum of 15 in. Hg.

For conveying filter aids (diatomaceous earth and perlite), a modified diaphragm liquid pump was developed by the combined efforts of Pacific Pneumatics Incorporated, Great Lakes Carbon Company, and the Gorman Rupp Company. This diagphragm pump will handle finely divided aeratable materials having bulk densities of 15–40 lb/cu ft. However, it will not handle materials that tend to stick together, such as paint pigments. Maximum discharge pressures run 5–7 psig.

When the materials are abrasive or not conducive to the passage through rotary feeders and the pressure reaches a maximum of 15 psig, the low-pressure blow tank and Airslide* pump (a form of blow tank) are applicable. The basis for the 15 psig maximum allowable pressure is the ASME pressure vessel code, upon which most state laws regarding the construction of pressure vessels are based. A vessel designed and constructed to operate above 15 psig must meet special design specifications and strict fabrication procedures. A vessel designed to operate at or below 15 psig must certainly be designed and constructed to withstand this internal pressure plus a factor of safety, but the stringent restrictions of the ASME code are not necessary. The blow tank and Airslide pump systems are intermittent in their operation. The can fill and empty in separate operations, but they cannot do both simultaneously. The blow tank system can handle both pulverized and granular materials. The Airslide pump can handle only pulverized materials.

The blow tank and Airslide pump feeders can and usually do have a holding hopper or head of material above them (Figure 2.8). With these units having intermittent operation, it is imperative to fill them as quickly as possible so that more time is available for emptying and conveying. This added time is reflected in a smaller pipeline system, since the conveying rate is minimized. The smaller pipeline system wil enable smaller-sized blowers and their driving motors to be used, as well as smaller-sized dust col-

---

* Registered trademark, Fuller Company, Bethlehem, Pennsylvania.

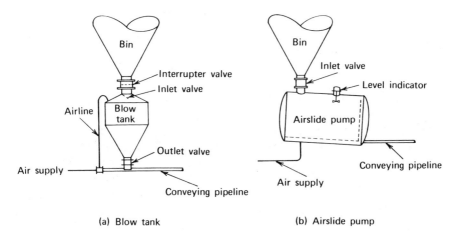

FIGURE 2.8.   (*a*) Blow tank and (*b*) airslide pump feeders.

lecting equipment at the terminal end, resulting in a more economical system in both original and operating costs.

Another type of system is the fluid solids pump for conveying over relatively short distances (maximum, 125 ft) and under intermittent or light duty. It is basically an adaptation of the heavy-duty fluid solids pump described in Section 2.3. It is used for bin and rail car unloading of pulverized materials. For bin unloading, its relatively low height—16 in. from bottom of base plate to top of hopper flange—minimizes head room. Between the bin outlet flange and the pump hopper, an open and shut valve should be used to shut off flow out of the bin. With the addition of a 90° short radius aerating elbow at the pump discharge, a vertical lift up to 40 ft is possible without an intermediate conveying distance. This provides, in effect, a "pneumatic elevator" for reclaiming from ground storage to adjacent process bins.

A development of the past 20 years is the Fox venturi eductor using low-pressure (4–10 psig) for conveying granular solids. Materials conveyed include talc, plastics, coal, limestone, ash, starch, catalysts, graphite, and cement. Pipeline sizes up to 6 in. are used, with the venturi nozzles sized according to system requirements.

2.3 MEDIUM-PRESSURE SYSTEM

The medium-pressure system is categorized as such by the air pressure requirement, 15–45 psig, the feeding mechanism, and the type of material it can convey. The system using the fluid solids pump normally requires that the material must be pulverized to a fineness having a minimum of 60% of the material passing through a 200 mesh screen, 75% passing through a 100 mesh screen, and all the material passing through a 50 mesh screen.

It is desirable to have 45% of the material pass through a 325 mesh screen to increase fluidization of the material. The material should be dry, free-flowing, and have particle shape and other characteristics that contribute to fluidization. The maximum conveying distance for this system is considered to be 4000 ft.

The original fluid solids pump is the Fuller-Kinyon* pump. This Fuller-Kinyon conveying system had more impact on the development of pneumatic conveyors than probably any other single piece of equipment. It was invented by Alonzo G. Kinyon just before World War I, with the first commercial installation being made in 1919. The original objective was to eliminate the fire and explosion hazard incident to the handling of pulverized fuels.

Kinyon conceived his invention while attempting to obtain uniform firing of a malleable melting furnace in a steel plant at Ashtabula, Ohio. Pulverized coal was delivered from the hopper of a pulverized coal bin by means of a triplex feeder. This feeder comprised three parallel screws having a common feed hopper. At times, the feed of fuel was intermittent, being either less than expected from the screw feeds or entirely interrupted. Following these conditions, the rate of feed exceeded that expected from the feeder speed, and sometimes the coal would flow even though the feeder was stopped. Kinyon discovered that the feeder undermined a material arch in the bin hopper. This collapsed either when a street car passed the plant or when a trip hammer shook the bin. At those times, the air below the arch mixed with the pulverized coal with the result that the coal tended to flow like a liquid. He verified this by filling a steel drum about half full of pulverized coal, which he aerated with a compressed air lance. He discovered that upon stirring the coal the apparent volume about doubled, at which time the aerated coal would flow as a liquid with a very slight angle of repose.

Kinyon assumed that if he could advance the coal into the pipeline system by screw pressure and introduce aerating air at a point beyond the end of the screw and a pressure seal or choke, the material could be transported through the pipeline by the continuous displacement of coal into it. More specifically, he felt that the screw pressure and displacement would be adequate with very little of the conveying pressure on the stream from the aerating air.

The importance of the conception and ultimate development of the Kinyon invention impressed engineering and scientific societies. On June 2, 1926, the Franklin Institute of Philadelphia awarded Alonzo G. Kinyon its Edward Longstreth Medal for his "New and Important Improvements in the Art of Conveying Pulverized Materials." Later in 1926, Kinyon was made a member of the Franklin Institute.

In the early development of the Fuller-Kinyon system, it was discovered that by increasing the total air volume, power requirements decreased and

* Trademark, Fuller Company, Bethlehem, Pennsylvania.

the flow became much more uniform. The early systems operating with high-pressure air actually depended on the kinetic energy of the expanding air. Air volumes were progressively increased so that very little of the conveying pressure is now attributed to the displacement effect of the screw. Many mechanical modifications were made, but the present arrangement retains the original principle of the differential pitch impeller screw to compress the material to form the positive and continuous seal against the system pressure.

With the application of the individual multivane rotary air compressor, the Fuller-Kinyon system was gradually converted to its present operating air volumes and lower pressures. The air volumes are higher, therefore in the handling of combustibles the ratio of air volume to the coal handled was increased, but it never exceeded 1% of the air necessary to support combustion.

From the original material conveyed, pulverized coal, the Fuller-Kinyon system is now conveying close to 40 different materials. The materials being conveyed include pulverized materials such as Portland cement, pulverized coal, limestone dust, flue dust, pulverized phosphate rock, and other materials of similar physical structure. The system is most commonly used for conveying from pulverizer and mill collecting conveyors serving a plurality of mills, from storage bins to packer bins, in-plant conveying, and for loading and unloading railroad cars, ships, and barges. The system is the one most commonly used by the cement industry for conveying raw mix and finished cements. It is and has been used for unloading and conveying cement on many major construction projects such as Hoover, Bonneville, and Grand Coulee dams, the St. Lawrence Seaway, and many large industrial structures. Its ability to handle abrasive materials is its greatest asset. The first known application of hard surfacing of metals to resist abrasion was on the impeller screw of the Fuller-Kinyon pump. Its inhibition to blowback of conveying air when properly operated minimizes the dust problem at the pump.

The basic system (Figure 2.9) comprises (1) the Fuller-Kinyon pump; (2) the conveying pipeline of Schedule 40 black steel pipe with diverter valves (3) transports the material from the pump to its destination. Level indicators (4) have a dual objective. They not only act as a sensor of the system to automatic operation but also indicate on the automatic control panel (5) the exact status of materials in the bins.

In car unloading, the pump is often installed in a pit below the track, allowing the material in the car to flow out through the influence of gravity directly into the pump hopper. The pump hopper is extended up with a fabricated steel plate section to a point level with the bottom of the rail. At this point, flexible connections are used to connect this section to the outlet gates of the covered hopper car through quick-connecting clamps. When no car is being unloaded, the openings, together with the flexible connections, are protected from the elements by a sheet metal cover. At all times,

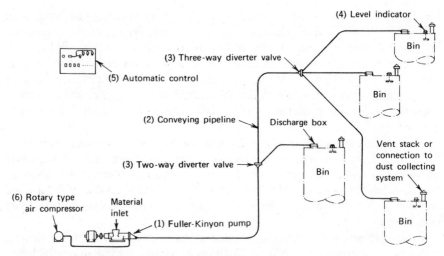

FIGURE 2.9.   Medium-pressure system using fluid solids pump.

the pump and its hopper must be protected against rain, hail, sleet, and snow.

To avoid construction of a pit under the track with its required supporting structure for the rails, the combining of an air-activated gravity conveyor with the pump allows the pump to be placed alongside the track or at some reasonable distance from it. For car connection, the same type of equipment and construction is employed as that for the pump directly below the track.

Adaptation of the pump to other than stationary units has enhanced its overall usefulness. Where a material or materials are to be conveyed from a row of vertical silos (Figure 2.10), the pump is mounted on a traveling carriage or a small version of a railroad flatcar. The carriage is equipped with flanged wheels, which are guided by the rails embedded in the floor. Flexible connections are used to connect the inlet material pump hopper to the open-and-close rotary valve at the outlet of the silo, as well as to the compressed air supply. The discharge outlet of the pump is connected to the conveying pipeline also through a flexible connection. This arrangement is most satisfactory when the unit is positioned for conveying over a period of several hours or more, since the movement of the pump from one silo outlet to another requires time for disconnecting and reconnecting.

Where frequent change of withdrawal from two or more silos is required, collecting conveyors, either mechanical, vibratory, or air activated, should transfer the materials from the silo outlets to a single pump. This will require additional headroom to accommodate the transfer conveyors, but it may minimize the capital expenditure as well as provide for greater operational flexibility.

Another adaptation, necessitated by the need to convey material from a flat area such as storage buildings or barge holds, is the mounting of the pump on a pair of individually powered rubber-tired wheels to make it entirely portable. Remote control of the pump allows the machine to be used without subjecting the operator to the hazard of working under unsafe conditions.

Although the pump impeller screw is balanced when new, after some period of operation, usually at standard 1200 rpm motor speed, the balance may become distorted through wear or other causes. To minimize wear of the screw and the barrel in which it rotates, material should always be available for conveying. The pump should not be allowed to operate for a sustained period without material or substantially underloaded.

The blow tank and Airslide pump system is also classified as a medium-pressure type system when the air requirements fall into the 15–45 psig range. They must, under such pressure conditions, be constructed according to the ASME pressure vessel code.

FIGURE 2.10. Movable fluid solids pump. (Courtesy Fuller Company, Bethlehem, Pennsylvania.)

2.4   HIGH-PRESSURE SYSTEM

High-pressure systems are those that use air pressure above 45 psig, reaching a maximum of 125 psig. The equipment choice is limited to the blow tank, which is especially well suited to long-distance conveying of pulverulent materials.

They were developed about 50 years ago by F. L. Smidth & Company. Their Fluxo* pumps were used in the construction of Hoover Dam, about 30 mi southeast of Las Vegas, Nevada, during the 1930s. Two systems were installed, each a two-tank system, and each with its own air supply, the first, to convey cement from the blending plant to a transfer station and the second from the transfer station to the mixing plant storage, both at a conveying rate of 450 barrels/hr. The first system delivered cement through a 9-in. ID pipeline 2600 ft long and a drop in elevation of 530 ft. The second system also used a 9-in. ID pipeline but was 3000 ft long and had a rise in elevation of 100 ft at the mixing plant.

The operation was so successful that after six months operation, bypassing the second pump and delivering the entire 6000 ft with one pump was tried. This too was successful with no appreciable decrease in conveying rate and no plugging of the pipeline. Some years later, similar equipment was used to convey cement for the construction of Grand Coulee Dam, Coulee Dam, Washington. This dual tank system conveyed cement over a distance of 7600 ft at a rate of 1000 barrels/hr using a single continuous 16-in. OD pipeline. Air pressure was 115 psig.

Two types of blow tank systems are manufactured: single tank and dual tank. The selection of the type and size as well as the transport pipeline is determined by the material and quantity to be transported over a given time, the length and figuration of the conveying pipeline, and the method of feeding. Air requirements (quantity and pressure) are determined by the system requirements.

The single tank is an intermittent system. It can only convey or transport material when it is discharging. It cannot fill and convey at the same time. To achieve a more continuous conveying cycle, the two-tank system is used. While one tank is discharging, the other tank is filling. By cycling the filling and discharging operation, conveying in the transport pipeline is very nearly continuous. The period of nonconveying is very slight, only that period of time necessary to switch from one tank to the other through automatic means.

For historical purposes only, these early single blow tanks are shown in Figure 2.11. Today's blow tanks have a vastly different configuration, as shown in Figure 2.12, a dual tank system. Instead of a top material outlet, bottom outlets are employed, which eliminates the air entry piping and valving at the bottom of the tank while adding the annular piping with air

---

* Registered trademark, F. L. Smidth & Company, Cresskill, New Jersey.

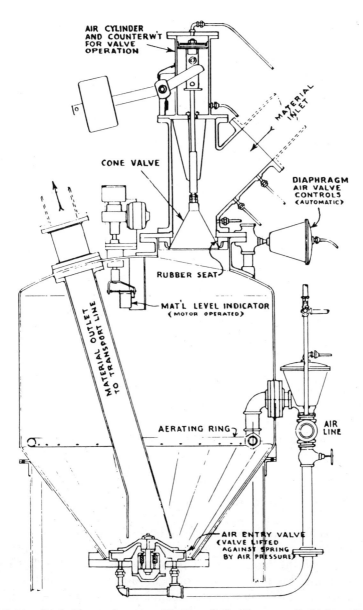

FIGURE 2.11. Single blow tank system. (Courtesy Fuller Company, Bethlehem, Pennsylvania.)

nozzles to homogenize the material completely for faster and cleaner discharge. Aeration is accomplished in the feed tank and aeration chute rather than in the original tank employing an aerating ring. This has lessened the time required for the tank to start discharging.

Another type of high-pressure (100 psig) blow tank is the ejector type for

FIGURE 2.12. Dual blow tank system. (Courtesy F. L. Smidth & Co., Cresskill, New Jersey.)

conveying sludges, semidried solids, wet slurries, grit, and screenings. These tanks are void of any internal aeration equipment such as that used on handling dry pulverulent materials. They rely on the power of expanding compressed air to discharge the material also through a bottom outlet into the conveying pipeline.

2.5 HIGH-DENSITY PULSE PHASE SYSTEM

The high-density pulse phase system can be described as a low- to medium-pressure (10–30 psig) blow tank system using air both to fluidize the material and to pressurize the blow tank for bottom outlet discharge, but with an added dimension, this added dimension being an air knife to alternately feed air and material into the conveying pipeline in slug fashion. The material moves through the pipeline not as a homogeneous mass mixed with air, but as cohesive masses separated by intervals of compressed air (Figure 2.13). The system was originally developed at the Warren Spring Laboratory of England's Department of Trade and Industry in the late 1960s and early 1970s.

Another method being employed in similarly operating systems is to maintain a continuous air flow with intermittent feeding or material into the

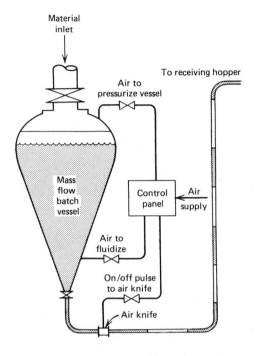

FIGURE 2.13. Pulse phase concept. (Courtesy Sturtevant Engineering Company, Ltd., Brighton, East Sussex, United Kingdom.)

conveying pipeline. This method is used successfully not only on low- and medium-pressure blow tank systems but also on high-pressure ones (90 psig).

For these systems to be successful, the material as it is conveyed through the pipeline should be able to remain in plug form. It should have high air permeability and low air retention.

2.6 LOW-PRESSURE, VENTURI FEED SYSTEMS

Normally, low-pressure, venturi feed systems are designed for operation below 2 psig using fans or turboblowers as the activators (Figure 2.14). By

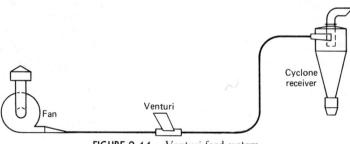

FIGURE 2.14. Venturi feed system.

the use of the venturi principle (i.e., a constriction in the pipeline prior to the material intake) causes the air velocity to increase to the point where the velocity pressure exceeds the static pressure, inducing a negative pressure at the material intake. The static pressure is the static resistance of the pipeline system during the conveying of material downstream from the intake point.

Only a metered feed of material to the system should be used. It cannot handle slugs or even a slight overcapacity. These will upset the venturi action, causing a no-flow or plugged condition in the conveying pipeline.

Some consider the venturi system the gentlest of all pneumatic conveyors. It has been used for frozen vegetables and some friable materials, but its most extensive application is the handling of plastic, paper and paperboard scrap, and continuous strips of paper or plastics, more commonly known as trim.

2.7 COMBINATION VACUUM–PRESSURE SYSTEM

The combination vacuum–pressure system is used when conveying from a multiplicity of pickup or originating points to a multiplicity of discharge or terminal points. It is also used where vacuum pickup is most desirable or mandatory and pressure delivery is most suitable. After discharge from the vacuum system, the pressure system can be any of the types—low, medium, or high—depending on the materials being conveyed and the distance to be traversed.

The original vacuum–pressure system is one that uses the positive pressure blower as a combination exhauster-blower (Figure 2.15). This system, however, has limited use, since the combined load, vacuum, and pressure across the blower is limited to the maximum allowable differential this machine is subjected to, normally 12 psig. Since the maximum vacuum used in that leg of the system is usually 12 in. Hg (6 lb for all practical purposes), the maximum pressure is limited to 6 psig. If higher pressures are required, the vacuum side would have to be designed to operate at lower vacuums, thus staying within the allowable differential across the blower.

To avoid this restriction, the two legs of the system are completely divorced, making the vacuum and pressure systems separate entities. They are, however, interlocked to prevent malfunction of either one in relation to the other. It is at this point that any type of pressure system could be used, though mostly low- and medium-pressure systems are used.

In coordinating these separate vacuum and pressure systems, a very critical point in their operation is that where the material leaves the vacuum zone and enters the pressure zone. Assume that a vacuum system and a low-pressure system are used with a single rotary airlock subjected to the vacuum above it and the pressure below it (Figure 2.16). The vacuum system is operating at a vacuum of 12 in. Hg. The pressure system is operating at

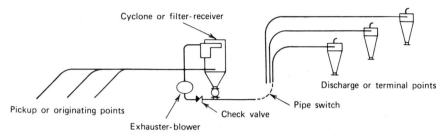

FIGURE 2.15. Combination vacuum-pressure system using a single blower.

10 psig. At these operating vacuums and pressures the differential across the feeder totals 16 lb. The leakage through the feeder from the pressure zone to the vacuum zone will be considerable. It will affect the required amount of air to be handled by both the vacuum system exhauster and the pressure system blower. But most important is the blasting effect of this blowback air into the filter-receiver. If the material being conveyed is either light or fluidizable under the influence of air, it will tend to hang up in the cone of the filter-receiver and not flow from the vacuum zone into the feeder pockets, causing a plugged filter-receiver. If the material does not flow, the blasting effect of the air from the feeder will also materially affect the efficiency and life of the filter media in the filter-receiver above.

In the handling of coarse material, this blasting effect may be alleviated by installing a blast deflector in the bottom cone of the filter-receiver (Figure 2.17a). This is an inverted cone placed above the outlet, but with sufficient clearance between its edge and the inside of the filter-receiver cone to allow material to drop toward the outlet. This deflector should be of extra-heavy construction with its supports out of the mainstream of material.

In the handling of fine, aeratable, and fluid materials, the blast deflector, while protecting the filter media above, will be an accessory to the buildup or hangup of material in the cone of the filter-receiver. To minimize this buildup, the differential across the feeder at the filter-receiver outlet must

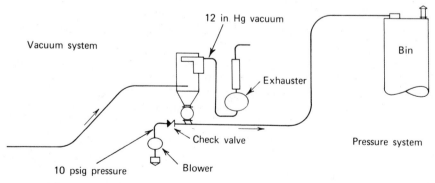

FIGURE 2.16. Combination vacuum-pressure system using separate blowers.

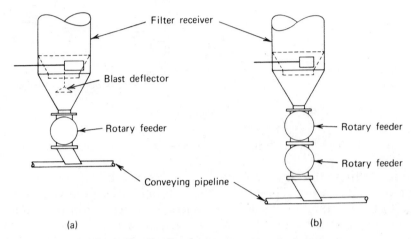

FIGURE 2.17.   Feeding from vacuum to pressure zones.

be reduced. The accepted method for this is to use two rotary feeders, one above the other. The top feeder should have a slightly higher volumetric displacement than the lower feeder (Figure 2.17*b*). This can be taken care of by a slightly higher speed of the rotor when both feeders are alike. Both feeders should operate in the same direction.

Where the material and amount to be conveyed are compatible to both the vacuum and fluid solids type of systems, this combination makes a most satisfactory installation. Each type is a separate entity connected by a surge hopper between the discharge of the vacuum system and the inlet to the fluid solids pump. The purpose of the surge hopper is to maintain a level of material so that the fluid solids pump has at all times material available to it. The surge hopper should be equipped with high, mean, and low level indicators. The high level indicator is interlocked with the vacuum system to either shut down that system or activate a vacuum breaker in the conveying pipeline to stop conveying when material reaches the level indicator to prevent backup of material in the system. The mean and low level indicators are interlocked with the fluid solids pump. The low level indicator should stop the pump when material is below the indicator and the mean indicator should start the pump when material reaches it. This arrangement will guarantee that material is always available to the pump. The air supply to the pump and conveying system must continue at all times. After conveying is completed, the remainder that is between the pump screw and the low level indicator can be conveyed by manually locking out the indicator to run the pump until the material is conveyed to its destination. A drop in air pressure requirements will indicate that the material has been conveyed and that the transport pipeline has cleaned out. It is at this point that the system is to be shut down.

In the handling of abrasive materials not suited to the fluid solids pump, a blow tank for the pressure leg is readily adaptable. The filter-receiver of the vacuum system is equipped with a larger than usual bottom cone, so that material can accumulate while the blow tank is performing its function of filling, emptying, and pressuring away the material it received from the vacuum system. Automatic controls allow the blow tank to return to a vacuum condition after emptying so that flow out of the filter-receiver into the blow tank is unimpeded. While conveying by the blow tank pressure system is intermittent, the vacuum system is continuous. To prevent overfilling the filter-receiver cone, a high-level indicator is installed at a point slightly below the lower level of the filter bags to shut down the vacuum system when material reaches its probe.

A single-unit vacuum–pressure system developed within the last 12 years is the Cyclonaire,* which uses the venturi-induced vacuum-load principle, loads the pressure tank by vacuum (15 in. Hg), and discharges by low pressure (15 psig) into the conveying pipeline. While the discharge of the pressure-type vessel is like all single-blow tanks on a batch principle, automatic control of its inlet cone valve and its discharge flap valve assures continuous operation.

In combination systems, several types can be used in conjunction with each other. Arrangement, purpose, and material being conveyed will have a direct bearing on the type to be used.

2.8 CLOSED-CIRCUIT SYSTEM

The closed-circuit system is ideally suited for conveying materials that are light and dry and whose dust would be hazardous; hygroscopic materials, which of necessity must be kept dry; materials that are hazardous because of an explosion factor, necessitating a blanket of inert gas; toxic materials; and materials of such value that losses must be minimized. The system uses a completely enclosed conveying circuit.

For low capacity (conveying rate) and short conveying distances, the centrifugal blower or turbofan (Figure 2.18) is used to activate the system. In such a system, material is usually fed into the pipeline system through a rotary feeder, with the material conveyed by the airstream to a cyclone receiver. In the cyclone receiver, the material is separated from the airstream and discharged through the airlock or rotary feeder at its bottom outlet. With the receiver having less than 100% separation efficiency, a small percentage is retained in the airstream to the inlet of the centrifugal blower or turbofan. The ability of this type blower or fan to pass dusts through them without damage to the wheels or housings makes them ideally suited. They are, however, limited by their inherent characteristic that for every speed and pressure of a centrifugal blower or turbofan there is a certain

* Registered trademark, Cyclonaire Corporation, Henderson, Nebraska.

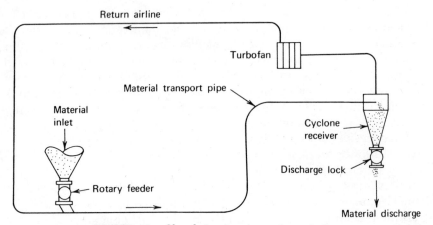

FIGURE 2.18.   Closed-circuit system using turbofan.

minimum volume below which the machine does not operate properly, causing its air or gas delivery to become irregular or unstable. Total operating pressures across the blower or fan seldom exceed 3 psig. Above this operating pressure, a rotary positive lobe blower is substituted. This unit in single stage can reach a differential pressure between its inlet and outlet of 12 psig. It is with this type of unit that higher conveying rates and longer distances can be achieved.

With the substitution of the rotary positive blower for the fan, it is now impossible to use only a cyclone receiver separator to separate the material totally from the conveying airstream. The rotary positive blower, having close clearances and tolerances between the impellers themselves and the impellers and the body and head plates, is unable to pass dust for any appreciable length of time under this service. The cyclone separator must be replaced with a cloth tube air–material separator to retain the dust and to prevent the dust from passing on to the blower. The cyclone separator can be retained, but the cloth tube air–material separator or filter (Figure 2.19) must be added between the air discharge of the cyclone separator and the inlet of the blower. This does, however, add another rotary feeder to the required equipment.

In years past, many in the industry thought closed-circuit systems, especially the cyclone type, were impractical. They thought that the system would fill up with material from the inefficiency of the cyclone separator. This is not so. In practice, the closed-circuit system stabilizes itself after a few moments of operation. When cyclones are properly designed, their efficiency rises quite sharply as the amount of material entering the cyclone increases. Usually the rate of efficiency increase is higher than the rate of increase in conveying.

In either the low- or high-capacity closed-circuit system, multiple intake

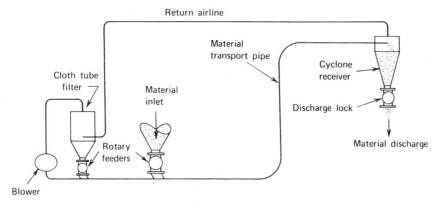

FIGURE 2.19.   Closed-circuit system using positive-pressure blower.

and discharge points are possible. Through the addition of conveying line and airline valves, various combinations can be used.

2.9   AIR-ACTIVATED GRAVITY CONVEYOR

Although not a pneumatic conveyor according to proper definition, the air-activated gravity conveyor is included as a member of the family. In a pneumatic conveyor, the motive force is supplied by the energy of the expanding airstream. In the air-activated gravity conveyor, gravity is the motive force. Normally, gravity conveying is thought of as involving material dropping down through a vertical or sloping spout whose angle from the horizontal is greater than the angle of repose of the material when that material is at rest or the angle at which the material will be piled up at the sides on a level surface after all the material that will freely slide has slid off.

All materials, whether coarse or finely divided, have a definite angle of repose. To convey material at a flatter angle than its angle of repose, some assist must be given the material. Vibration has been known for years as one of the best methods of conveying until a near level condition in material travel is met. It is at this point that the many horizontal conveyors came into use, including the air-activated gravity conveyor.

In the use of the air-activated gravity conveyor, the physical characteristics of the material to be conveyed determine the necessary air requirements and angle of inclination or, indeed, whether this device can be used at all. The material must be dry and aeratable. To be aeratable, the material should be finely divided, normally with a minimum of about 40% of the material able to pass through a 200 mesh sieve. Under the influence of air, the material should become fluidized so that it flows like water, seeking its own level.

Figure 2.20 illustrates the action of this conveyor, assuming the material to be cement having an angle of repose ranging from 47 to 52°. It is easy to visualize the angle of repose of a powdered material and its effect on the movement of that material. First imagine a hollow square framework of wood or metal that is supported at its edges in a level position. Imagine that over the top of this frame is stretched a heavy, porous fabric having a uniform and controlled permeability, or capacity to pass air under a given set of conditions.

By piling cement on top of this mounted fabric, or diffuser area, we would find that when the cement started to run off at all sides, the cone or pyramid remaining would have an angle of from 47 to 52°. If we wished to make all of the cement slide off the frame, it would have to be tipped up to an angle of 47° or slightly more for gravity to have sufficient effect on the cement to make it slide completely from the tilted surface.

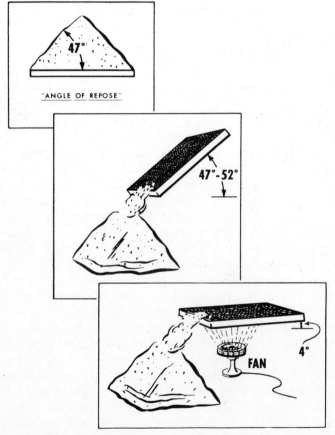

FIGURE 2.20.   Action of the air-activated gravity conveyor. (Courtesy Freuhauf Division, Freuhauf Corporation, Detroit, Michigan.)

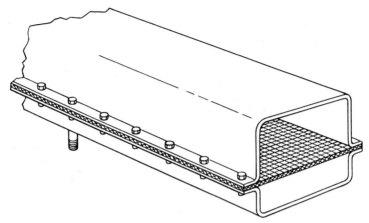

FIGURE 2.21.   Enclosed air-activated gravity conveyor. (Courtesy Fuller Company, Bethlehem, Pennsylvania.)

Returning the frame to its flat position, replacing the cement on top of the mounted fabric to the same 47° angle of repose, and passing air up through the porous fabric, we find the frame will have to be tilted only about 4° to clear the cement completely off the fabric. What actually happens is that the cement becomes aerated, removing the angle of repose at the base of the material. In short, the material is fluidized by means of air blown into it through the porous fabric. When the panel is placed at the proper angle, the aerated material on it is caused to flow to the lowest point of the panel, seeking its own level, much like water in a river flowing from its source in the mountains to its destination, the ocean.

To determine whether or not an air-activated gravity conveyor will operate, certain tests should be made. It is known that many materials having the same name and chemical composition do not necessarily have the same flow characteristics.

It is essential to determine the aerated angle of repose of the material to be conveyed. To avoid errors, such as guessing or making erroneous implications, the material should be tested on equipment suitable to the purpose. This equipment includes an air-activated gravity conveyor capable of being adjusted to various slopes, roughly level to 15° from the horizontal. A gauge indicating the slope should be attached to the conveyor and readily readable. For measuring the air supply, a manometer and a flowmeter are used simultaneously to ascertain proper conveying conditions together with the proper slope of the conveyor. The result of these tests give the designer the minimum slope required as well as optimum slope to use, and the proper air supply, based on pressure and volume required.

This conveyor, like the pneumatic conveyor pipeline, has the ability to go around corners, using a long radius curved section for the change in direction. The slope, however, must be maintained.

For conveying, the section above the porous medium is enclosed to contain the material as it travels from the inlet to the discharge point (Figure 2.21). Attachments to the conveyor include gates to regulate the flow on the conveyor, side discharge gates to discharge material at any given point in the conveyor, and accurate cutoff for scale feeding operations. This conveyor can also be used to split the mainstream into two or more streams. Since the material is fluid as it moves down the conveyor, it will have a uniform depth across the conveyor, which should be level, as well as near uniform speed. Although this is not a 100% perfect split, it is reasonable to consider it so for many purposes.

Several porous media may be used. The most prominent are the woven cotton and polyester fabric and the porous plates made from grains of fused crystalline aluminum oxide, silica sand, silicon carbide, and other materials. The woven fabric, although it resembles woven conveyor belting to the naked eye, is a fabric woven under exacting manufacturing conditions to have uniform and controlled permeability throughout its entire length and breadth. A fabric having a nonuniform permeability will allow air to pass up through the cloth in an erratic manner, giving uneven aeration to the material and hence producing unstable conveying conditions. The porous plates are also manufactured under rigid conditions to meet the same requirements. Other materials in use include porous plastic and rubber products.

Beside conveying, this type of conveyor is used to discharge material from bins and to aerate material in the bin to accelerate the flow of material out of the bin. One of the early applications to bin unloading and discharge was horizontal bulk flour bins. This proved so successful that today most bulk flour bins, either vertical or horizontal, use this method of withdrawal. Aeration of material within a cone-bottom bin is accomplished by using open-top conveyors, three or more installed radially against the inside of the bottom cone. Aerating or fluidizing the material reduces surface friction between the material and the bin's internal surfaces; this will tend to break down arching to allow the material to discharge from the bin.

3 DESIGN OF PNEUMATIC CONVEYORS

Pneumatic conveying, being the trickiest of all the material-handling arts, presents a real problem in systems design. Although theory has been applied in an attempt to arrive at a foolproof set of design formulas, the result has been nebulous because of the inability to translate to a theoretical formula what happens in actual practice. The many variables of material density, particle size and shape, moisture content, arrangement of the conveying pipeline, especially the number and relative location of pipe bends, air density, and varying amounts of air and materials all combine to give a limitless variety to any one or number of factors. Any formula for the design of a pneumatic conveyor must therefore be compromised to allow for these many variables. It cannot distinguish between heat or cold, dry or humid conditions or adverse climate. Common sense and experience are necessary. A farm commentator several years ago ended his presentation with a bit of advice that seems appropriate: "Learn from the experience of others. You won't live long enough to make all of the mistakes yourself."

The information in this chapter is based on experience only. This is not a total guarantee against pitfalls, but from the author's experience, when this information is used with understanding and restraint, results are fairly reliable.

3.1 CRITERIA FOR DESIGN AND USE

Before we can arrive at what we need to know to design and apply pneumatic conveyors, it must be remembered that a pneumatic conveyor is not the solution to all conveying problems. There is most definitely a place in the bulk material handling field for the mechanical conveyors, i.e., screw,

37

bucket, belt; the vibrating type of conveyors; the front-end unloader; and even the shovel and wheelbarrow. Many times a combination of these and pneumatic conveyors is used to solve bulk conveying problems. The purpose of the conveyor, its arrangement, and its basic requirement should be accurately determined and applied to the function of the operation required. The material to be conveyed should be known by all its characteristics, including density, particle size and shape, temperature, susceptibility to moisture, fragility, and abrasiveness.

The density of the material is its weight per cubic foot. The weight should be known under four classifications: in a poured condition, tapped to settle and condense the material, stored in a bin, and when under the influence of aeration. Materials being conveyed in a pneumatic conveyor are handled many times under all four of these classifications. The material may enter the system in the poured state, but when it is discharged, it may be in the aerated state, which means that more bulk is involved at the discharge than at the entrance. Volumetric capacity, as well as flowability, is vitally affected by this phenomenon.

The first step in determining the particle size of the material is visual inspection. From this visual inspection, the material can be classified under four general terms: very fine, fine, granular, and lumpy or irregular. Very fine materials are those that will pass through a 100 mesh sieve. Fine materials will all pass through a $\frac{1}{8}$ in. mesh, or a 6 mesh sieve. Granular materials will all pass through a $\frac{1}{2}$ in. mesh screen. Lumpy materials are those having particles greater than $\frac{1}{2}$ in. in any direction, and irregular materials are those that are fibrous, stringy, or the like. If the materials are either very fine or fine, a more accurate determination should be made by a sieve analysis. This sieve analysis should always be given in terms of either the Tyler Standard Screen Scale as developed by C. E. Tyler (W. S. Tyler, Inc., Mentor, Ohio) or the U.S. Standard Sieve Series of the National Bureaus of Standards, Washington, D.C. They are both universally used.

The moisture content of the material must be known. Its importance can be noted in a material such as paper makers clay. The conveying rate of systems handling this material at 0–3% moisture will be reduced about 15% for each additional percentage of moisture. Not only must the moisture content be known, but so must the susceptibility of the material to moisture (hygroscopicness). The degree to which it is hygroscopic will affect the conveying rate as well as its suitability for conveying. If the material is hygroscopic and readily absorbs moisture, does it deliquesce? A deliquescent material, if it is conveyable, may require dry air, or other medium, for conveying.

The pH, a symbol denoting acidity or alkalinity of material, should be known. pH values run from 0 to 14. A pH of 7 indicates neutrality. Numbers less than 7 indicate increasing acidity and numbers greater than 7 denote increasing alkalinity. The pH of the material to be conveyed reflects very much the type of cloth to use in filter-receivers and dust-retentive devices.

There are materials known by a common name whose pHs vary all over the scale. One such material is salt cake (impure sodium sulfate). Laboratory tests have shown this material to have a pH of 3.2–9.1. Unless the pH is known, the designer can only guess what type of cloth to use in filter-receivers and dust collectors.

The corrosiveness of the material should be known. The pH of the material may determine this characteristic, but it is best to spell out the degree to which the material will corrode various metals. A highly corrosive material may require special metals of construction.

The explosive or combustible nature of the material or its dust must be known. Even though pneumatic conveyors are known to be among the safest techniques in handling such materials, precautions in the system must be taken. Some materials may require inert gas to be used as the conveying medium. Other materials may require that special metals be used to avoid sparking within the components of the system, and where appreciable volumes are prevalent, explosion vents or hatches might be required.

Is the material fragile? Will attrition to the particle size during conveying be detrimental to the end use of the material? An answer to this question is necessary. Many materials of the prilled type are susceptible to breakage in a pneumatic conveyor. Materials are prilled for many reasons, mainly elimination of dust and ease of handling. If the prills are broken during conveying, irreparable damage is done, including the ruination of the material for its intended end use. Seed grains and similar materials must be handled with care. Velocity within the system, if not controlled, can be a most discriminating factor.

The material's abrasiveness should be known since this characteristic, as much as any other, will determine the type of system and equipment to use. Using the Moh scale for hardness to indicate abrasiveness in pneumatic conveyors, a generalization of materials classification might be made as follows:

Moh Scale	Classification
1	Nonabrasive
2	Nonabrasive
3	Slightly abrasive
4	Medium abrasive
5	Medium abrasive
6	Highly abrasive
7	Highly abrasive

Materials having a hardness of above 7 are for the most part too abrasive to be handled in pneumatic conveyors. Putting it simply, when metal is to be cleaned, abrasives are used at normally high velocities. Each time the action takes place, metal is eroded and thicknesses reduced. Translating this action to a pipeline, high velocities to the interiors of cyclones, and rotating parts inside closely machined housings, rapid wear of the equipment results in short life. The more universal use of the blow tank together with chrome–nickel alloy spun-cast pipe, however, has made pneumatic conveyors applicable for handling many abrasive materials.

After determining everything about the material to be conveyed, a tabulation in concise form should be made for ready reference. Such a tabulation is shown in Table 3.1.

While the material will determine to a great extent the type of pneumatic conveyor to use, the purpose to which it is applied together with the rate of conveying will also determine the type. As much as possible should therefore be known about where the material comes from and where it is to go.

TABLE 3.1. Material Specifications

MATERIAL: Common Name_____ Chemical Name_____

WEIGHT PER CUBIC FOOT: Poured_____ Tapped_____
                        Stored_____ Aerated_____

ANGLE OF REPOSE: _____

PARTICLE SIZE: (give classification and range of size of particle)
                Lumpy or Irregular_____Range_____
                Granular_____Range_____
                Fine_____Very fine_____
                (If fine or very fine, give sieve analysis below)

Screen Size Mesh	Weight on or between Sieves		Total Percentage	
	Grams	Percent	On	Passing
8	—	—	—	—
14	—	—	—	—
20	—	—	—	—
35	—	—	—	—
48	—	—	—	—
65	—	—	—	—
100	—	—	—	—
200	—	—	—	—
325	—	—	—	—
Pan	—	—	—	—

MOISTURE CONTENT_____ TEMPERATURE_____

SPECIAL CHARACTERISTICS: pH_____ Toxic_____ Fragile_____
   Abrasive_____ Hygroscopic_____ Deliquescent_____
   Corrosive_____ Combustible_____ Explosive_____

Where the material comes from will determine the type of feed to the pneumatic conveyor. Is the material at rest, or is it in motion? At rest, the material would be in a hopper, bin, flat storage area, or some type of railroad car, barge, ship, or truck. In motion, the material would be discharging from other material-handling or process equipment. If the material is free-flowing and has sufficient velocity as it leaves hoppers or bins, it can be also considered in motion.

If the material is at rest, force must be applied to accelerate the material in the pipeline conveying system. If it is in motion as it enters the pipeline conveying system, less force is required for acceleration, hence lower operating pressures or vacuums are required with resulting reduced power requirements.

For railroad car unloading, the type of car used to transport the material over the rails should be indicated: boxcar, standard covered hopper car, or one of the specialty cars. In the case of boxcar, it must be remembered that the material is at rest, will not move unless disturbed, and will not come out by itself. We must go into the car with a hose and nozzle connected to a vacuum-type system to get the material out. Covered hopper cars are normally used for materials that are relatively free flowing or that require a medium amount of vibration or aeration to the material to get it to flow out of the discharge gates at the bottom of the car. The pneumatic conveyor can only convey the material from the discharge outlet, it cannot exert its influence into the car to enhance the flow characteristics of the material in the car. There are exceptions to this when sufficient air is entrained in the material to satisfy the conveying requirements and when the material is sufficiently free flowing that external inducement to flow is not required.

Where the material is received by the pneumatic conveyor in motion from other material-handling devices or process equipment, rates of flow must be accurately determined. In such cases, the pneumatic conveyor must be designed with a built-in factor of safety to take away not only the material delivered to it but also the surges that may occur as well as possible conservative ratings attached to the material-handling devices or process equipment. The timing of the delivery, whether it is continuous or intermittent, also should be known. If continuous, conveying is a simple matter. If intermittent, care and caution must be observed. Under this condition, also known as batching, the volume of each batch must be known as well as the frequency at which each batch is discharged to the pneumatic conveyor. The conveyor should be sized so that each batch is completely conveyed prior to the acceptance of the next batch. This prevents backup of material before the entry of material into the conveyor. The rate of conveying is now determined by process rates or by the speed at which railroad cars, trucks, or ships are to be either loaded or unloaded.

In conjunction with conveying rates, the severity of operation should also be determined. The construction of a pneumatic conveyor can be likened to the degree of design and construction of speed reducers for various applications as developed by the American Gear Manufacturers Association. With the more sophisticated uses to which pneumatic conveyors are put, their operating frequency is becoming more diverse. It is only natural to assume that equipment designed to operate at periodic intervals would not stand up under continuous operation. Severity of operation can be classified as follows:

Class	Maximum Hours of Operation
1	20 hours per week
2	40 hours per week
3	80 hours per week
4	160 hours per week

With this classification, the degree of sturdiness to which the conveyor must be constructed can be determined. It will also tell the appurtenances necessary to meet the requirements of the operation.

The type and number of receptacles to which the pneumatic conveyor will discharge must be known. If there is more than one, their relative location to each other should be known both in plan and elevation. After determining the location and type of feed to the pneumatic conveyor as well as the discharge points, the path of the conveying pipeline must be found. Since the most advantageous pipeline layout is the shortest distance from the feed point to the discharge (a straight line), it is imperative that this straight-line approach be maintained as much as possible. The number of changes in direction should always be kept to a minimum. When change in direction, horizontal or vertical or a combination of the two, is necessary, long radius bends must be used. For nonabrasive and mildly abrasive materials, these bends should have a radius equal to 12 in. for each inch in diameter of the conveying pipeline. For abrasive materials requiring chrome–nickel alloy cast pipe bends, the radius usually used is 3 in. for each inch in diameter of the conveying pipeline.

If more than one long radius bend is required in the pipeline, they should not be run concurrently. One bend right after another in a conveying pipeline can only result in a deficient system. As the material is conveyed around a bend, the centrifugal force tends to transpose the material from the total area of the pipe to the back of the bend. The added friction of material against the wall of the pipe will slow down the velocity of the material as it traverses the bend. If the bends were in immediate succession, this slowing down of the material would drop out of suspension in the airstream, causing the pipeline to plug. A good rule of thumb is to have a length of straight pipe between the point of tangency of the first bend and the point of curvature of the second bend at least equal to 40 times the inside diameter of the pipe. This will allow the material to accelerate sufficiently after going through the first bend to successfully get around the succeeding bend.

TABLE 3.2. Purpose and Conveying Rate

TYPE OF OPERATION: (describe)_____

_____

_____

_____

From _____ To _____

DISTANCE: Horizontal_____ Vertical_____ No. of Bends_____
(For accurate layout of conveying pipeline indicate on a sketch distances in
plan and elevation together with obstructions to be avoided and clearances to
be observed. Also show space available for equipment.)

CONVEYING RATE: Per hour_____ Per day_____

SEVERITY OF OPERATION: Class_____

TYPE OF CONVEYING: Continuous_____ Intermittent_____

ENVIRONMENT: Atmosphere_____ Temperature_____
Hazardous Dust_____ Hazardous Gas_____

ELEVATION (ft above sea level)_____

Elevation of the installation in relation to sea level must always be known. Atmospheric pressure varies with differences in altitude. At sea level and 70°F temperature the atmospheric pressure is 14.7 psi, whereas at 3000 ft elevation it is 13.16 psi, and at 6000 ft elevation it is 11.77 psi. Difference in air density is very pronounced in the operation of pneumatic conveyors and should always be considered.

The environment in which the equipment is to be operated should be known. Will it be subjected to normal atmospheric conditions and temperatures, or will other conditions occur such as high or low temperatures, hazardous dusts and gases, or corrosive elements? These conditions will determine materials of construction and type of equipment such as electric motor versus air motor operation of appurtenances.

In addition to the tabulation of the material characteristics or specifications (Table 3.1), a tabulation of purpose and conveying rate should be made (Table 3.2).

Completion of all parts of tabulations of material characteristics and purpose and conveying rates will give all the information required to determine the type of system to use and provide information needed for the design.

3.2 DESIGN PROCEDURES

Once all the information about the material to be conveyed, physical arrangement, and requirements has been gathered, the first decision is the type of system to use. This is based on the type of material to be conveyed and purpose of the system.

The type of system to use is first determined by the particle size of the material. A quick determination can be made from Table 3.3.

Another type of determination is made by the characteristics of the material other than particle size. Deviations from Table 3.3 may be made, since only general conditions have been used. Special conditions of operation may call for the unexpected, which cannot be isolated in a general tabulation. System type can be determined from the material characteristics (Table 3.4).

Table 3.5 indicates the various types of conveyor applicable to the many known materials. Systems 2.5 (high-density pulse phase) and 2.6 (low-pressure venturi feed), have been omitted. These systems are very dependant on uniformity of physical characteristics of the materials to be conveyed. (This is especially true of the high-density pulse phase.) Inclusion could lead to erroneous applications.

Tables 3.3, 3.4, and 3.5 are excellent guides. They are by no means, however, the ultimate answer. The amount of material to be conveyed, the length of the conveying pipeline, and the arrangement of the conveying system may necessitate deviations from the recommendations indicated. It is at this point where the art in pneumatic conveying begins to be meaningful.

After the type system is resolved, the necessary air requirements to move the material through the pipeline system are determined. Limitations now imposed limit design by formulas to vacuum and low-pressure systems. Medium- and high-pressure systems are devoid of design by formulas. The basis for design of such systems is actual experience in the handling of the same or similar materials over comparable distances and conveying rates. Such systems operate at close to absolute minimum velocities. The air-to-material ratios will vary because of minute differences in physical characteristics, mainly particle size and the ability of the material to be fluidized or aerated.

Vacuum systems operate under negative pressures up to 12 in. Hg. Beyond this negative pressure the air loses its impinging and carrying characteristics. In low-pressure systems up to 12 psig, the impinging and carrying characteristics can be reliably predicted, hence such systems can be closely designed. In any design, however, the unpredictable factor of how the material acts under the influence of air must be given consideration. This factor can be gleaned only from experience.

The design of a pneumatic conveyor can be likened to the chef preparing a tossed salad to satisfy the gustatory sense. No matter how good the quality of the main ingredients, the little things like herbs and garnishes determine the savoriness. In the conveyor, while the power and air requirements may be theoretically and practically correct, the type, choice, and layout of the various components, the salt and pepper so to speak, will determine the success of the system.

In determining the power and air required, together with the operating vacuum and pressure, several differing approaches have been made. The

TABLE 3.3. System Type According to Particle Size

Particle Size	Type of System						
	2.1, Vacuum	2.2, Low-Pressure	2.3, Medium-Pressure	2.4, High-Pressure	2.7, Vacuum-Pressure	2.8, Closed-Circuit	2.9, Air-Activated
Lumpy or irregular	x	x					
Granular	x	x		x	x	x	
Fine	x	x		x	x	x	?
Very fine	x	x	x	x	x	x	x

TABLE 3.4. System Type According to Material Characteristics

Material Characteristic	Type of System						
	2.1, Vacuum	2.2, Low-Pressure	2.3, Medium-Pressure	2.4, High-Pressure	2.7, Vacuum-Pressure	2.8, Closed-Circuit	2.9, Air-Activated
Nonabrasive	x	x	x	x	x	x	x
Abrasive, slightly	x	x	x	x	x	x	x
Abrasive, medium	?	?	x	x	?		x
Abrasive, highly	?	?	x	x			
Hydroscopic	x	x	x	x		x	?
Deliquescent	?	?	?	?	?	x	?
Combustible	x	x	x		x	x	x
Explosive	x	x	x			x	
pH (acid)	x	x	x	x	x	x	x
pH (alkaline)	x	x	x	x	x	x	x
Toxic						x	
Fragile	x	x	?			x	x
Corrosive	?	?	?	?	?	?	?
Thermoplastic	x	x			?	x	x

46

TABLE 3.5. System Type According to Known Material

Material	Type of System						
	2.1 Vacuum	2.2 Low-Pressure	2.3 Medium-Pressure	2.4 High-Pressure	2.7 Vacuum-Pressure	2.8 Closed-Circuit	2.9 Air-Activated
Adipic acid	x	x			x	x	?
Alfalfa meal	x	x			x	x	?
Alum	x	x			x		
Alumina, floury	x		x	x	x		x
Alumina, sandy	x			x			x
Aluminum hydrate	x	x	x		x		x
Aluminum oxide	x		x	x			x
Ammonium sulfate	x	x					
Arsenic oxide	x	x				x	
Asbestos	x					x	
Barley	x	x			x		
Barytes	x		x				x
Bauxite	x		x				x
Beet pulp, dried	x	x				x	
Bentonite	x	x	x		x		x
Bone meal	x	x					
Borax	x	x	x	x			?
Boric acid	x	x					x
Bran	x	x					
Brewers dried grains	x	x					
Calcium carbonate	x	x	x		x		x
Calcium phosphate			x				
Carbon, activated	x	x	x		x	x	x

47

TABLE 3.5. (Continued)

Material	Type of System						
	2.1 Vacuum	2.2 Low-Pressure	2.3 Medium-Pressure	2.4 High-Pressure	2.7 Vacuum-Pressure	2.8 Closed-Circuit	2.9 Air-Activated
Carbon black, pelletized	?	?					
Catalysts, petroleum	x	x	x		x		
Cellulose acetate	x	x			x	x	?
Cement, Portland	x		x	x			x
Cement, raw materials	x		x				x
Cereals	x	x			x		
Cerelose	x	x			x		x
Chalk	x	x	x	x	x		x
Citrus pulp, dried	x	x				x	
Clay, kaolin	x	x	x		x		?
Coal, pulverized		x	x	x			x
Coal, sized anthracite		x					
Coal, slack bituminous		x					
Cocoa, powdered	x	x				x	x
Coffee beans	x	x			x		
Coke, fines	x	x					
Coke, flour	x	x	x				x
Copra	x	x					
Corn	x	x			x		
Corn grits	x	x			x		
Cottonseeds, delinted	x	x			x		
Cottonseed meal	x	x			x		
Detergent powders	x	x	x		x		

Material							
Diatomaceous earth	x		x		x		x
Distillers grains, dried	x				x		
Dolomite	x		x		x		x
Feed ingredients	x			x	x	x	
Feeds, soft	x				x	x	
Feldspar	x		x				x
Ferric sulfate	x						
Ferrous sulfate	x						
Fertilizers	?		?	?			?
Fish meal	x				x	x	
Flaxseed	x				x		
Flint	x		x				x
Flour, wheat	x			x	x		x
Fluorspar			x				x
Fly ash	x		x	x			x
Fullers earth	x		x		x		x
Gluten meal	x				x		
Grain, ground	x				x	x	
Grain, whole	x				x		
Graphite	x				x	x	
Grits, brewers, coarse	x				x		
Grits, brewers, refined	x				x		
Gypsum	x		x		x		x
Hog fuel	x				x		?
Ilmenite ore			x				
Kaolin	x		x		x		x
Lime, hydrate	x			x	x	x	x
Lime, pebble	x			x	x		x
Lime, lump	x				x		
Limestone, pulverized			x				x
Linseed oil meal	x				x		

49

TABLE 3.5. (Continued)

Material	2.1 Vacuum	2.2 Low-Pressure	2.3 Medium-Pressure	2.4 High-Pressure	2.7 Vacuum-Pressure	2.8 Closed-Circuit	2.9 Air-Activated
Magnesium oxide	x				x		?
Malt, brewers	x	x					
Malt, brewers, ground	x	x			x		
Marble dust			x				x
Meat scraps	x	x		x			
Milk, powdered	x	x					
Nylon, pelleted	x	x			x		x
Nylon, powdered	x	x			x		
Oats	x	x			x		
Oat flour	x	x					
Ores, pulverized			x	x			?
Peanuts, whole	?	?					x
Perlite	x	x			x		x
Petroleum coke	x	x	x				x
Phosphate rock, pulverized			x	x			?
Plaster		x	?				
Polyethylene pellets	x	x			x		x
Polyethylene, powdered	x	x			x		x
Polyvinyl chloride	x	x			x		x
Pulp, wood	x	x					
Pyrites			x	x			
Resins	x	x			x	x	x
Rice	x	x			x		x
Rubber, crumb	x	x				x	
Rubber, pelletized	x	x					
Rye	x	x					

Type of System

50

Material							
Salt	x				x		
Salt cake	x				x		
Sand	x		x	x	x		?
Sawdust	x	x					
Seeds	x						
Semolina	x						
Silica, pulverized		x	x	x	x		x
Soap ingredients	x		?		x		?
Soda, anhydrous, caustic	x	x					
Soda ash, light	x	x			x		x
Soda ash, dense	?	?					
Sodium carbonate	x	x	x		x		x
Sodium phosphates	x	x	x		x		x
Sodium sulfate	x	x			x		?
Sodium sulfite	x	x					
Sorghum grains (kafir and milo)	x	x					
Soybeans	x	x					
Soybean meal	x	x					
Starch, corn, pelletized	x	x					
Starch, corn, pulverized	x	x	x		x		x
Starch, potato	x						
Sugar, granulated	x	x			x		x
Sugar, powdered	x	x					x
Talc	x	x	x		x		x
Titanium dioxide	?	?	x				?
Urea, prilled	?	?					
Wheat	x	x			x		
Whiting	x	x	x	x	x		x
Wood chips	?	x					
Wood flour	x	x					x
Zinc oxide	x	x	x			x	

experience approach makes use of two factors, saturation (cubic feet of air required to convey 1 pound of material) and hp/ton (power required to convey 1 ton of material in 1 hour). In compiling these two factors, as well as their relationship to each other, it was found that the operating vacuum of the system is a factor related to the ratio of horsepower per ton to the saturation. Low ratios indicate low vacuums and high ratios indicate high vacuums. This further shows that the denser the conveying stream (more material per cubic foot of air), the higher the vacuum.

A constant operating vacuum for all systems and materials may induce conveying characteristics detrimental to the material and the overall operation of the system. At the intake to the system, a certain amount of air and energy is required to accelerate the material into the pipeline system. At this point of the system, the air is drawn in at 14.7 psia. As energy is applied to move the material, the energy is reflected in a partial vacuum at the intake to the exhauster. Since the air as it moves through the pipeline is rarified, its volume becomes greater. As the volume increases, so does the velocity within the pipeline. This velocity increase, plus the air–material mixture, is reflected in higher frictional resistance, adding to the partial vacuum at the intake to the exhauster. In long-distance conveying, it is necessary at times to "step" the pipe diameter in the system to stabilize the operating vacuum and velocity within operable limits.

The design of vacuum systems on a saturation basis is determined by the material to be conveyed, rate of conveying, and length of the conveying pipeline. The figures given in the saturation table (Table 3.6) include average curvature (120°/100 ft of pipeline), discharge lock loss below the air–material separator, and air-reversal loss in the separator itself. If the curvature exceeds the amount included in the tabulated figures, add 6 in. to the length of the pipeline for each additional degree of curvature.

If the system is used to unload railroad cars, either boxcar or covered hopper, the saturation figures given in the table are for the use of smooth bore flexible metal hose in connecting the unloading nozzles to the stationary pipeline. For 5-in. and under conveying pipeline systems, the hose will be the same diameter as the pipeline. In such systems, it is wise to increase the velocity about 10 per cent to give a little added inertia to the airstream for material acceleration. For a 6-in. pipeline system, 5-in. hose is used, a 7-in. pipeline system uses a 6-in. hose, and for an 8-in. pipeline system, either two 5-in. hoses or one 7-in. hose is used. In larger systems, the net area of the hoses should be about 10 to 15 per cent less than the area of the pipeline.

For nonabrasive and mildly abrasive materials, the flexible metal hose with hardened or stainless steel overlapping liner will give satisfactory performance and maintenance life. When abrasive materials are handled, flexible rubber hoses having smooth bore and wound static wires are used. Since rubber has a higher frictional resistance to the passage of air and material than steel, we can expect reduced conveying rates for comparably sized

TABLE 3.6.   Saturation—Vacuum System

| Material | Wt per cu ft | Conveying Distance | | | | | | | | Velocity ft/sec |
| | | 100 ft | | 150 ft | | 250 ft | | 400 ft | | |
		Sat.	hp/T	Sat.	hp/T	Sat.	hp/T	Sat.	hp/T	
Alum	50	3.6	4.5	3.9	5.0	4.3	5.7	4.7	6.3	110
Alumina	60	2.4	4.0	2.8	4.7	3.4	5.7	4.0	6.4	105
Carbonate, calcium	25–30	3.1	4.2	3.6	5.0	3.9	5.5	4.2	6.0	110
Cellulose acetate	22	3.2	4.7	3.5	5.1	3.8	5.7	4.1	6.0	100
Clay, air floated	30	3.3	4.5	3.5	5.0	3.9	5.5	4.2	6.0	105
Clay, water washed	40–50	3.5	5.0	3.8	5.6	4.2	6.5	4.5	7.2	115
Clay, spray dried	60	3.4	4.7	3.6	5.2	4.0	6.2	4.4	7.1	110
Coffee beans	42	1.2	2.0	1.6	3.0	2.1	3.5	2.4	4.2	75
Corn, shelled	45	1.9	2.5	2.1	2.9	2.4	3.6	2.8	4.3	105
Flour, wheat	40	1.5	3.0	1.7	3.3	2.0	3.7	2.5	4.4	90
Grits, corn	33	1.7	2.5	2.2	3.0	2.9	4.0	3.5	4.8	100
Lime, pebble	56	2.8	3.8	3.0	4.0	3.4	4.7	3.9	5.4	105
Lime, hydrated	30	2.1	3.3	2.4	3.9	2.8	4.7	3.4	6.0	90
Malt	28	1.8	2.5	2.0	2.8	2.3	3.4	2.8	4.2	100
Oats	25	2.3	3.0	2.6	3.5	3.0	4.4	3.4	5.2	100
Phosphate, trisodium	65	3.1	4.2	3.6	5.0	3.9	5.5	4.2	6.0	110
Polyethylene pellets	30	1.2	2.0	1.6	3.0	2.1	3.5	2.4	4.2	80
Rubber pellets	40	2.9	4.2	3.5	5.0	4.0	6.0	4.5	7.2	110
Salt cake	90	4.0	6.5	4.2	6.8	4.6	7.5	5.0	8.5	120
Soda ash, light	35	3.1	4.2	3.6	5.0	3.9	5.5	4.2	6.0	110
Soft feeds	20–40	3.0	4.2	3.4	4.5	3.7	5.0	4.2	5.5	110
Starch, pulverized	40	1.7	3.0	2.0	3.4	2.6	4.0	3.4	5.0	90
Sugar, granulated	50	3.0	3.7	3.2	4.0	3.4	5.2	3.9	6.0	110
Wheat	48	1.9	2.5	2.1	2.9	2.4	3.6	2.8	4.3	105
Wood flour	12–20	2.5	3.5	2.8	4.0	3.4	4.9	4.4	6.5	100

hoses. On systems using 25–30 ft of hose, the conveying rate will be reduced by about 10% when using rubber hoses instead of metal.

A step-by-step outline of the procedure follows.

1.   From Table 3.6 for the material and conveying distance, determine the saturation (cubic feet of free air per pound of material to be conveyed per minute) and the hp/ton (horsepower required to convey 1 ton of material in a period of 1 hour). The saturation figures in the table are for 4-, 5-, and 6-in. ID conveying pipes. Since frictional resistance varies not only by the

square of the velocity but also by pipe diameter, larger pipelines (when conveying rate requirements demand) use slightly lower saturations. For 8-in. conveying pipe, saturation can be reduced 15% and the hp/ton 15%. For 10-in. conveying pipe, saturation can be reduced 25%, and for 12-in. conveying pipe, 35%. Horsepower per ton factors to use with these saturation factors are most difficult to determine, so such systems should be designed for an operating vacuum of 12 in. Hg.

For conveying distances longer than 400 ft, extreme caution must be used. Saturations must be increased, but on a sliding scale. For a 550-ft conveying distance, increase the saturation factor for 400 ft by 17%; a 700-ft conveying distance, by 30%; an 850-ft conveying distance, by 41%; and a 1000-ft conveying distance, by 50%.

Further corrections may be made, depending on the feed to the conveying system. The saturation factors are for a nonmetered feed, or a material at rest. For metered feeds, or where material is in motion as it enters the conveying pipeline, saturation factors may be reduced 10–15%, depending on the rate and flowability of the material. If there is any doubt regarding the motion of the material, it is better to disregard this correction.

For preliminary determination of the pipeline size, we can assume that 4, 5, and 6 in. can handle a conveying rate of up to 12 tons/hr; 8 in., up to 25 tons/hr; and 10 in., up to 40 tons/hr.

2.   The volume of free air required for the system (scfm) is computed by multiplying the saturation determined in step 1 by the conveying rate (in pounds per minute):

$$\text{scfm} = \text{saturation} \times \text{conveying rate (pounds per minute)}$$

3.   The size of the conveying pipe required is next determined by computing the constant of that particular pipe by dividing the scfm from step 2 by the velocity as given in the saturation table (Table 3.6):

$$\text{pipe constant} = \frac{\text{scfm}}{\text{velocity (fps)}}$$

Constants for various pipe diameters are given in Table 3.7.

If the pipe constant falls between pipe sizes, use the larger pipe size and recalculate the scfm by multiplying the constant of the larger pipe by the velocity.

When the pipeline exceeds 12 in., its size can be determined by calculating the internal area required to handle the air requirement at the desired velocity by the following formula:

$$\text{area (sq in.)} = \frac{\text{cfm} \times 144}{\text{velocity (fps)} \times 60}$$

TABLE 3.7. Pipe Constants

IPS Pipe Size	Schedule	Pipe Constant			
		5	10	30	40
3 in.		3.6	3.5		3.07
3½ in.		4.8	4.6		4.05
4 in.		6.1	5.9		5.3
5 in.		9.4	9.2		8.4
6 in.		13.5	13.2		12.0
7 in.					16.0
8 in.		23.2	22.7	21.3	
10 in.				34.0	
12 in.				47.8	

4. The vacuum at which the system will operate when conveying at its rated capacity is determined from a factor that is arrived at by dividing the hp/ton factor by the saturation factor determined in step 1:

$$\text{vacuum factor} = \frac{\text{hp/ton}}{\text{saturation}}$$

When this vacuum factor is below 1.3, the operating vacuum will be 8 in. Hg. When it is between 1.3 and 1.4, the vacuum will be 9 in. Hg; between 1.4 and 1.5, 10 in. Hg; between 1.5 and 1.6, 11 in. Hg; and above 1.6, 12 in. Hg.

5. With the scfm required by the system, and the vacuum at which the system will operate, we can now size the blower needed to activate the system. This is determined by the actual amount of air the unit is to inhale. The actual amount of air, acfm, is expanded air at the intake conditions to the blower, which in a vacuum system is used as a vacuum pump or exhauster. This acfm is calculated by multiplying the scfm by 30 in. Hg divided by 30 in. Hg minus the operating vacuum:

$$\text{acfm} = \frac{\text{scfm} \times 30}{30 - \text{operating vacuum}}$$

6. Blower size may now be determined by consulting the various manufacturer's catalogs, choosing the size in accordance with the acfm and the operating vacuum. It is recommended that blowers in pneumatic conveying service operate at 15% below the maximum operating speed recommended by the manufacturer. Then, in the event the blower must be speeded up after operation gets under way, it can be done and the speed will still be under the maximum allowed. It is also recommended that choice of the type of blower be based on the severity of the operation.

7.   The required speed of the blower is determined by dividing the acfm by the blower displacement in cubic feet per revolution, plus a slip allowance. The slip allowance, expressed in rpm, has been determined by all manufacturers for each of their type and gear diameters.

$$\text{rpm} = \frac{\text{acfm}}{\text{blower displacement}} + \text{slip allowance}$$

If this rpm is greater than the maximum recommended speed less 15%, the next larger-sized blower should be used.

8.   The horsepower required to drive the blower is calculated by the following formula:

$$\text{hp} = \text{rpm} \times \text{displacement(cf/r)} \times \frac{\text{vacuum}}{2} \times 0.005$$

For approximate horsepower requirements, the following formula can be used:

$$\text{hp(approx.)} = \text{acfm} \times 1.20 \times \frac{\text{vacuum}}{2} \times 0.005$$

All the preceding formulas and calculations have been based on sea level conditions of 29.92 in. of mercury absolute pressure and 70°F. At elevated altitudes, the weight of air is reduced; thus to approximate the same conveying conditions there as at sea level, corrections are necessary to the scfm determined in step 2, while the same pipeline size used at sea level is retained within limits. The impinging effect of air at elevated altitudes is equal to the square root of the ratio of the altitude's absolute pressure to the absolute pressure at sea level. The correction factor table (Table 3.8) shows the factor by which the scfm at sea level is increased to provide the volume of air required at the higher elevation.

The correction factor for altitude is also to be applied to the blower slip when determining blower size.

As an illustration, consider pebble lime, ¾ in. and under, having a bulk density of 56 lb/cu ft, which is required at the rate of 500 tons/week for a manufacturing plant. The source of supply is 425 miles away, and the altitude of the plant is 250 ft above sea level.

1.   After studying transportation costs, we find that the most economical transport method is rail in covered hopper cars. At the plant, the storage bin is to be placed at the point of usage, 180 ft from the unloading track. Management has decided that a minimum supply on hand at all times should be that amount required for one week's operation, 500 tons. Since the bin

TABLE 3.8.   Atmospheric Pressure and Correction Factor at Various Altitudes

Altitude (feet above sea level)	P, Absolute Pressure (psi)	H, Absolute Pressure (in. Hg)	R, Correction Factor at that Altitude
0	14.69	29.92	1.00
1,000	14.16	28.86	1.02
2,000	13.66	27.82	1.04
3,000	13.16	26.81	1.055
4,000	12.68	25.84	1.08
5,000	12.22	24.89	1.095
6,000	11.77	23.98	1.12
7,000	11.33	23.09	1.14
8,000	10.91	22.22	1.16
9,000	10.50	21.38	1.18
10,000	10.10	20.58	1.20
11,000	9.71	19.75	1.23
12,000	9.34	19.03	1.25
13,000	8.97	18.29	1.28
14,000	8.62	17.57	1.30
15,000	8.28	16.88	1.33

is not full at all times, a 20% allowance is made for unbalanced deliveries with respect to supply and demand, increasing the size of the bin to 600 ton capacity. In conveying from cars, the most applicable system from both original equipment and operating cost is the straight vacuum type when delivering to one bin. After deciding to use this type, the filter-receiver is placed on top of the bin, making the total conveying distance 250 ft, including the 180-ft horizontal distance and an estimated figure of 70 ft vertically to the filter-receiver intake.

In the plant operation, it is desired to unload this lime within a 5-day, 40-hr week without incurring overtime. The lime will be received in 60-ton capacity covered hopper cars, with possibly boxcars being substituted in the event of a shortage of covered hoppers. Eight cars plus per week, or an average of two cars per day, are to be unloaded for a total of 120 tons/day. Two cars per day should be unloaded to avoid demurrage costs.

To determine the required conveying rate, allowance should be made for getting the cars ready for unloading and for maintenance and lubrication of the equipment, and, finally, for the time consumed in coffee breaks and other nonproductive periods outlined in the labor contract. All of these allowances amount to about 40% of the total required time. This may sound high, but it is a realistic figure. The conveying rate can now be set for the system itself by dividing the 120 tons/day by 8 hr/day to get a rate per hour of 15 tons. To get the actual rate of the system, since it will operate 60%

of the total time, we divide the 15 tons/hr by 0.6 and arrive at a good, not too liberal or conservative rate of 25 tons/hr.

With the rule-of-thumb method of determining the pipeline size, we find that for this rate of 25 tons/hr, an 8-in. pipe is required. From the saturation table, we find that 3.4 cu ft of air is required to convey 1 lb of lime over the distance of 250 ft. Since the saturation table is for 4-, 5-, and 6-in. pipelines, a correction in the required saturation is made for the larger 8-in. pipeline. This correction amounts to a reduction in the saturation figure of 15%. The corrected saturation is then 3.4 − (3.4 × 0.15), or 2.9. Since the material will be flowing out of a covered hopper car into the conveying system, the material can be considered in motion when it is received by the conveying pipeline. While this flow on pebble lime might be considered sluggish, a further reduction in the saturation figure of 10% can be made. The final and design saturation figure will be the corrected saturation of 2.9 for the 8-in. pipeline system, minus 10%, or 2.6. The hp/ton will remain 4.7. We can now proceed with the design.

2.
$$\text{scfm} = \frac{2.6 \times 50.000 \text{ lb/hr}}{60} = 2160$$

3.
$$\text{pipe constant} = \frac{2160 \text{ scfm}}{105 \text{ fps}} = 20.6$$

This pipe constant of 20.6 is slightly below the constant of 21.3 for an 8-in. schedule 30 pipe. In order to maintain the proper velocity in the conveying pipeline, the volume of free air (scfm) will be increased to 105 fps times the pipe constant, 21.3, to 2240 scfm. Corrected saturation will be 2240 scfm/833 lb/min, or 2.7.

4.
$$\text{operating vacuum} = \frac{4.7 - 15 \text{ per cent}}{2.7} = 1.48$$

System will operate at 10 in. Hg vacuum.

5. To find acfm at blower intake:

$$\frac{2240 \text{ scfm} \times 30 \text{ in. Hg}}{(30 \text{ in. Hg} - 10 \text{ in. Hg})} = 3360$$

6. Next we find the blower size. But first an explanation. All blower manufacturers list the performance data of their blowers on standard air at the inlet. (Standard consists of a temperature of 70°F, ambient pressure of 14.7 psia, and specific gravity of 1.0.) Blowers in vacuum service operate under different conditions, namely, ambient pressure of 14.7 psia minus

the amount of vacuum under which the system will operate—in this case, 9.7 psia.

Some manufacturers do and some do not list the volumetric capacity of the blower in cubic feet per revolution (cf/r). This figure should be known so that slip in revolutions can be determined. Further, slippage must be corrected for vacuum, since the inlet air at 9.7 psia is less dense than that at 14.7. Slippage from the performance data at standard conditions must therefore be corrected by multiplying this slip at equivalent pressure by a factor. This factor F is listed as follows (Table 3.9):

TABLE 3.9.   Vacuum Slippage Factor for Positive-Pressure Blower

Vacuum (in. Hg)	Equivalent Pressure (psig)	F
6	3	1.105
7	3.5	1.125
8	4	1.145
9	4.5	1.168
10	5	1.190
11	5.5	1.216
12	6	1.241

Now we can proceed. To adhere to the recommendation that blowers in pneumatic conveying service operate at no greater speed than 85% of the manufacturer's allowable recommendation, we divide the acfm at blower intake by 0.85 to determine the size from the performance data tables:

$$\frac{3360 \text{ acfm}}{0.85} = 3953$$

This will give us the required performance cfm of the blower from one of the many blower manufacturer's catalog data. Referring to catalog data, we find a size 1021 (10-in. gear diameter and 21-in. impeller length) has a full rated pressure of 9.0 psig, a maximum cfm of 4524 at 1770 rpm and 6.0 psig, and a displacement of 2.8 cf/r.

7. To find the speed we require, subtract from the 1770 rpm the rpm required to reduce the cfm capacity from 4524 to 3360 (from step 5) by their difference divided by the blower displacement:

$$1770 \text{ rpm} - \frac{(4524 \text{ cfm} - 3360)}{2.8 \text{ cf/r}} = 1354$$

To this must be added the additional slippage due to the blower operating under a vacuum. To find the slippage, multiply the rpm by the cf/r to get the cfm without slippage; then substrate the tabulated cfm from the cfm without slippage divided by the cf/r as follows:

$$2.8 \text{ cf/r} \times 1770 \text{ rpm} = 4956 \text{ cfm}$$

$$\frac{(4956 \text{ cfm} - 4524)}{2.8} = 154 \text{ rpm}$$

For slippage due to vacuum, multiply this rpm by the factor F and subtract this rpm:

$$(154 \text{ rpm} \times 1.190) - 154 = 30 \text{ rpm}$$

Total rpm is then 1354 + 30, or 1384.

8.   Horsepower required to drive the blower:

$$1384 \text{ rpm} \times 2.8 \text{ cf/r} \times 5.0 \text{ psig} \times .005 = 97 \text{ bhp}$$

Let us now move the plant in toto to an altitude of 4000 ft. At this altitude the absolute pressure is 12.68 psi and the correction factor is 1.08. The rule-of-thumb choice of pipeline size is still 8 in., and the basic saturation will remain at 2.6 cu ft of air per pound of material conveyed by this computation. The computations must now include the altitude correction factor as follows:

1. $$\text{scfm} = \frac{2.6 \times 50,000 \text{ lb/hr}}{60} \times 1.08 = 2335 \text{ cfm}$$

2. $$\text{pipe constant} = \frac{2335 \text{ cfm}}{105 \text{ f/s}} = 22.2$$

This pipe constant 22.2 is slightly above the constant of 21.3 for an 8-in., schedule 30 pipe. For all practical purposes we will continue to use the 8-in. pipe and the air requirements (scfm) calculated in (2). Although this will increase the velocity within the pipeline to 110 ft/sec, the frictional resistance within the pipeline will remain about the same as that at sea level, since lighter air induces less resistance.

3. $$\text{operating vacuum} = \frac{4.7}{2.9} = 1.48 \text{ (10 in. Hg)}$$

4.   acfm at blower intake $$= \frac{2335 \text{ cfm} \times 25.84 \text{ in. Hg}}{(25.84 \text{ in. Hg} - 10.00 \text{ in. Hg})} = 3815 \text{ cfm}$$

5. For blower size, first divide the acfm by 0.85:

$$\frac{3815 \text{ acfm}}{0.85} = 4488$$

We find that we can use the same size blower, 1021, since its maximum cfm is 4524.

6. Blower speed:

$$1770 \text{ rpm} - \frac{4524 \text{ cfm} - 3815}{2.85 \text{ cf/r}} = 1517 \text{ rpm}$$

In addition, we must add the slippage due to the higher elevation of 4000 ft by multiplying the slip at elevation 250 ft by the R factor in Table 3.8 as follows:

$$(154 \text{ rpm} \times 1.190 \times 1.08) - 154 = 44 \text{ rpm}$$

Total rpm is then 1517 + 44, or 1561.

7. Horsepower required to drive the blower:

$$1561 \text{ rpm} \times 2.8 \text{ cf/r} \times 5.0 \text{ psig} \times 0.005 = 109 \text{ bhp}$$

The blower should be driven by a 125-hp motor. In determining the size motor, do not forget to take into consideration the motor manufacturer's recommendations for higher elevations.

As the vacuum system is designed from saturation and horsepower per ton factors, so too is the low-pressure system (6–12 psig). A table of such factors has been developed (Table 3.10). Operating pressure many times is limited by the ability of the feeding device to feed material into the pipeline system against a positive pressure. Uniform feed will require a certain pressure and velocity, whereas intermittent feed for the same volume of material will require a higher velocity to accelerate the material, with possibly the same or lower pressure, but larger pipeline size. We know very little about the behavior of materials as they are inserted into the pressure pipeline system having a velocity sufficient to accelerate the material and convey it to the discharge end. Manufacturers and suppliers of pneumatic conveyor systems and components have developed guidelines compatible with their own type of equipment and experience. Hence preliminary design of power, air requirements, and pipeline size can be developed by the uninitiated, but the experience factor must be added. This experience factor is the ability to spot where deviations are necessary. It is not included in the procedures

TABLE 3.10.   Saturation—Low-Pressure System

Material	Wt. per cu ft	Pressure Factor	Conveying Distance						Velocity (ft/sec)
			100 ft		250 ft		400 ft		
			Sat.	hp/T	Sat.	hp/T	Sat.	hp/T	
Alum	50	4.0	1.6	2.7	2.0	3.4	2.2	3.8	65
Alumina	60	5.0	1.1	2.4	1.6	3.4	1.9	3.9	60
Carbonate, calcium	25–30	3.5	1.4	2.5	1.8	3.3	2.0	3.6	65
Cellulose acetate	22	3.0	1.4	2.8	1.7	3.4	1.9	3.6	55
Clay, air floated	30	4.0	1.5	2.7	1.8	3.3	1.9	3.6	50
Clay, spray dried	60	4.3	1.5	2.8	1.8	3.7	2.0	4.3	55
Clay, water washed	40–50	4.5	1.6	3.0	1.9	3.9	2.1	4.4	60
Coffee beans	42	5.0	0.6	1.2	0.9	2.1	1.1	2.5	45
Corn, shelled	45	5.0	0.9	1.5	1.1	2.2	1.3	2.6	55
Flour, wheat	40	2.5	0.7	1.8	0.9	2.2	1.1	2.7	35
Grits, corn, coarse	33	3.5	0.8	1.5	1.3	2.4	1.6	2.9	70
Lime, pebble	56	5.0	1.3	2.3	1.6	2.8	1.8	3.3	70
Lime, hydrated	25–30	4.0	0.6	1.8	0.8	2.2	0.9	2.6	40
Malt	28	5.0	0.8	1.5	1.1	2.0	1.3	2.5	55
Oats	25	5.0	1.0	1.8	1.4	2.6	1.6	3.1	55
Phosphate, trisodium	65	4.5	1.4	2.5	1.8	3.3	1.9	3.6	75
Polyethylene, pellets	30	5.0	0.55	1.2	0.9	2.1	1.1	2.5	70
Salt cake	90	5.0	2.9	3.9	3.5	4.5	4.0	5.1	83
Soda ash, light	35	5.0	1.4	2.5	1.8	3.3	1.9	3.6	65
Soft feeds	20–40	3.8	1.3	2.5	1.7	3.1	1.9	3.7	70
Starch, pulverized	40	3.0	0.8	1.7	1.1	2.4	1.5	3.0	55
Sugar, granulated	50	5.0	1.4	2.2	1.6	3.1	1.7	3.6	60
Wheat	48	5.0	0.9	1.5	1.1	2.1	1.3	2.6	55

that follow, since this experience factor is a multiple one, varied, and, at times, extremely isolated.

For design of low-pressure systems, the following steps are taken:

1.   From Table 3.10 for the material and conveying distance determine the saturation (cubic feet of air per pound of material to be conveyed per minute), and the hp/ton (horsepower required to convey 1 ton of material in a period of 1 hr). The saturation figures in the table are generally for 4-, 5-, and 6-in. ID conveying pipes. Since frictional resistance varies not only by the square of the velocity, but also by pipe diameter, larger pipelines may

accommodate a denser conveying stream at corresponding pressures and velocities. Conversely, smaller pipelines may require a less dense conveying stream, also at corresponding pressures and velocities. Since we are dealing with various pressure ratios, we will base our design on the one set of factors, and resolve the small discrepancies in this approach with further computations.

Unlike the vacuum system, there is no rule-of-thumb method for arriving at pipeline sizes for different conveying rates. Pipeline sizes must be determined from air requirements of the system and the velocity of the air in a compressed state at the feeding mechanism.

The author's experience on low-pressure systems longer than 600 ft is not as great as is his experience with vacuum systems. Thus, for want of more reliable information for the basis of possible corrections, these must be left to the discretion of the individual. In such decisions, try not to be too conservative. If the pipeline decided on is found to be too small to accommodate the conveying rate, no change in power or air delivery can offset the error made. Only a larger system is the answer.

2. The volume of free air required for the system (scfm) is computed by multiplying the saturation determined in step 1 by the conveying rate in pounds per minute:

$$\text{scfm} = \text{saturation} \times \text{conveying rate (lb/min)}$$

3. The operating pressure is next determined by dividing the hp/ton by the saturation and multiplying by the pressure factor:

$$\text{pressure (psig)} = \frac{\text{hp/ton}}{\text{saturation}} \times \text{pressure factor}$$

4. Since the velocity of the air in the saturation table (Table 3.10) is based on acfm at the point of feed, this acfm is computed by multiplying the scfm by the atmospheric pressure divided by the absolute pressure:

$$\text{acfm} = \frac{\text{scfm} \times 14.7}{14.7 + \text{pressure (psig)}}$$

5. The size of the conveying pipeline required is determined by the constant of that particular pipeline by dividing the acfm from step 4 by the velocity as given in the saturation table (Table 3.10):

$$\text{pipe constant} = \frac{\text{acfm}}{\text{velocity (ft/sec)}}$$

(For pipe constants for the various sizes of pipes, see Table 3.7.)

Up to this point, we have determined the required conveying air, operating pressure, and the pipeline size for conveying. To properly size the blower for the required conveying air, we must take into consideration the air loss through the feeding mechanism. Gate locks having tight seating gates will pass through them the volume of one compartment of the lock multiplied by the number of times per minute the lock is discharged. Rotary feeders have not only their volumetric displacement multiplied by the number of revolutions per minute the rotor operates, but also the leakage that occurs between the rotor blades and the body. This leakage will vary for each type of construction. A good rule of thumb is to increase the volumetric leakage by 30% to arrive at the total leakage involved. This may seem like a high figure to use, especially with a brand new feeder having very close clearances. It does, however, include compensation for wear that will allow a longer period of operation before maintenance to the feeder is required.

6.   Rotary feeder leakage is computed in acfm by multiplying the volumetric displacement by the revolutions per minute of the rotor and 1.3:

$$\text{feeder leakage (acfm)} = \text{volumetric displacement} \times \text{rpm} \times 1.3$$

7.   Since the air at the point in the system where the feeder operates is equal to or possibly a little higher than the operating pressure of the system, we must calculate the scfm required at the blower inlet:

$$\text{feeder leakage (scfm)} = \text{acfm (6)} \times \frac{14.7 + \text{pressure}}{14.7}$$

8.   Air delivery requirement of the blower is the sum total of the scfm required by the system and the feeder leakage:

$$\text{scfm (blower)} = \text{scfm (system)} + \text{scfm (feeder leakage)}$$

9.   Blower size may now be determined by consulting the various manufacturer's catalogs, choosing the size in accordance with the required scfm and the operating pressure. The blower speed, as recommended under the vacuum system computations, should apply for pressure systems too, i.e., to operate at 15% below the maximum operating speed recommended by the manufacturer. Also, choice of the type of blower should be based on the severity of operation.

10.   The required speed of the blower is determined by dividing the scfm by the blower displacement in cubic feet per revolution, plus a slip allowance. The slip allowance, expressed in rpm, has been determined by all manufacturers for each of their type and gear diameter. It is well to note that most manufacturers in their performance data give the capacities and horsepower based on standard air at the inlet to the blower. Standard air

is at 14.7 psia and 70°F. By interpolation between the performance data given in the catalogs the speed can be determined. Otherwise,

$$\text{rpm (blower)} = \frac{\text{scfm}}{\text{blower displacement}} + \text{slip}$$

11.   The horsepower needed to drive the blower can be determined from the manufacturer's performance data or is calculated by the following formula:

$$\text{hp} = \text{blower rpm} \times \text{displacement (cf/r)} \times \text{pressure (psig)} \times 0.005$$

In our calculations we may run into a situation where the original operating pressure from the saturation table may be considered too high for the ability of the feeder to feed the material into the pipeline stream, or the feeder itself cannot stand the high differential across it. In cases such as this, operating pressures can be reduced by increasing the saturation figure, which will result in more cfm in the system. With this incease in cfm, pipeline size must also be increased to maintain proper velocity and to reduce the overall frictional resistance that will allow operation at the reduced pressure. Velocities are noted in the saturation table should be maintained, no matter what the pressure, as long as the pressure remains within the range of 6–12 psig.

Having gone through the necessary steps of the design computations, let us now see how they are applied. First, a system is required to convey hydrated lime from four storage bins to a loading-out bin over a combined vertical and horizontal distance of 400 ft. The hydrated lime in the storage bins will have a bulk density of 22 lb/cu ft. The desired conveying rate is 60,000 lb/hr. As a matter of interest, a vacuum system would require something larger than an 8-in. pipeline and approaching 200 hp. Further, a large filter-receiver would be required at the terminal end of the system to retain the fine lime dust. For a pressure system, a single pickup point is preferable, so screw feeders and screw conveyors will be used to bring the lime from the outlets of the bins to the single rotary feeder pickup point or feeding mechanism. Hydrated lime, being a material susceptible to arching in the bottom cones of the storage bins, will require either mechanical or air assistance to flow with regularity. In this case, we will use open top Airslides to assist the flow out of the bins.

1.   From the saturation table (Table 3.10), we determine the saturation to be 0.9 and the hp/ton to be 2.6.

2.   From the saturation table (Table 3.10), we also determine the operating pressure:

$$\frac{2.6}{0.9} \times 4 = 11.6 \text{ psig}$$

3. For free air at this operating pressure, we calculate:

$$1000 \text{ lb/min} \times 0.9 = 900 \text{ scfm}$$

4. For acfm at the feeder, we calculate:

$$900 \text{ scfm} \times \frac{14.7}{(14.7 + 11.6)} = 508 \text{ acfm}$$

5. $$\text{pipe constant} = \frac{508 \text{ acfm}}{40 \text{ fps}} = 12.7$$

For this pipe constant, even though it is slightly higher than a 6-in., schedule 40, we can safely use this size pipe.

In determining the feeder size required, the lime in the bin has a bulk density of 22 lb/cu ft. In accelerating the flow of the lime out of the bin, however, aeration is used, which reduces its bulk density. The screw feeders and conveyors can also reduce bulk density through the turbulence within the conveyors. In the aerated condition, we can assume this lime to then have a bulk density of 18 lb/cu ft. Transposing the pounds per minute to be conveyed to cubic feet per minute, we find that the feeder must be capable of feeding 56 cu ft of material per minute from the discharge of the screw conveyors into the pipeline system. We decide to use a heavy-duty rotary feeder that can withstand the differential across it of 11.6 psig, a volumetric capacity of 3.5 cut ft/revolution of the rotor, and an air assist to force the lime out of the pockets of the rotary feeder into the pipeline system. Since the screw feeders should be designed as the metering feed, the rotary feeder will be used as an airlock only. Under this condition, the feeder should rotate at an efficiency of about 75%, or 22 rpm.

6. feeder leakage (acfm) = 3.5 cfr × 22 rpm × 1.3 = 100 cfm

7. feeder leakage (scfm) = 100 cfm × $\dfrac{(14.7 + 11.6 \text{ psig})}{14.7}$ = 179 cfm

8. blower air = 900 cfm + 179 = 1079 scfm

9. Power required is 76 bhp.

For illustrative reasons, suppose we wish to operate this system using a different type of feeder that would only allow us to operate at a maximum of 8 psig. It has been found that within certain tolerances the factors in the saturation table (Table 3.10) can be altered to reflect a change in operating pressure, which will also reflect changes in the saturation and the pipeline size. The maximum deviation allowed with the factors given in the table is a pressure that is 33% above or below that as derived from the factors. The reduction of 3.6 psig is within this allowance, so we are on fairly safe ground. The saturation at this reduced pressure of 8 psig is determined by multiplying the saturation at the original pressure by the square of the ratio of the original absolute pressure to the new absolute pressure:

$$\text{saturation at 8 psig} = 0.9 \times \left(\frac{26.3}{22.7}\right)^2 = 1.2$$

$$\text{scfm at 8 psig} = 1000 \text{ lb/min} \times 1.2 = 1200 \text{ cfm}$$

$$\text{acfm at feeder} = 1200 \text{ cfm} \times \frac{14.7}{(14.7 + 8)} = 780 \text{ cfm}$$

Since the velocity within the pipeline should not change, we again use the figure of 40 ft/sec. The pipe constant then is

$$\frac{780 \text{ cfm}}{40 \text{ pfs}} = 19.6$$

This falls between a 7- and an 8-in. pipeline. The 8-in. pipeline should be used, and since we are at about the maximum deviation in pressure from that in the saturation table, we will design the system around the 8-in. pipeline and an operating pressure of 8 psig. Our computations then evolve around these two operating parameters:

$$\text{acfm} = 21.3 \text{ (pipe constant)} \times 40 \text{ fps} = 852$$

$$\text{scfm} = 852 \text{ cfm} \times \frac{(14.7 + 8)}{14.7} = 1340$$

Assuming the same volumetric capacity in the feeder to be used here as in the 11.6 psig system, its leakage in acfm will be:

$$3.5 \text{ cfr} \times \frac{(14.7 + 8)}{14.7} \times 22 \text{ rpm} \times 1.3 = 155 \text{ cfm}$$

$$\text{air requirement for blower} = 1340 \text{ cfm} + 155 = 1495 \text{ cfm}$$

The power required at 8 psig will be 71 bhp.

For lower original cost of the conveying system (smaller pipeline size) and reduction in the size of the dust suppression equipment at the terminal end of the system due to less air being handled, the 6-in. system appears to be the proper one to use.

Several more illustrations with notations follow to give added insight into this low-pressure conveying system design and its ramifications.

For salt cake, weighing 90 lb/cu ft, is to be conveyed over a distance of 150 ft at the rate of 6 tons/hour:

1. Use saturation of 3.1, hp/ton of 4.1, and a pressure factor of 5.

2. Operating pressure $= \dfrac{4.1}{3.1} \times 5 = 6.6$ psig.

3. scfm $= 3.1 \times 200$ lb/min $= 620$ cfm.

4. acfm $= 620$ cfm $\times \dfrac{14.7}{(14.7 + 6.6)} = 429$ cfm.

5. Pipe constant $= 429$ cfm/83 fps $= 5.15$.

A 4-in. pipeline with an operating pressure of 6.6 psig is to be used.

For salt cake, weighing 90 lb/cu ft, is to be conveyed over a distance of 180 ft at the rate of 25 tons/hr:

1. Use saturation of 3.2, hp/ton of 4.2, and a pressure factor of 5. Since the conveying distance of 180 ft falls between 100 and 250 ft, interpolation of the factors for the two distances given is necessary.

2. Operating pressure $= \dfrac{4.2}{3.2} \times 5 = 6.6$ psig.

3. scfm $= 3.2 \times 833$ lb/min $= 2660$ cfm.

4. acfm $= 2660$ cfm $\times \dfrac{14.7}{(14.7 + 6.6)} = 1800$ cfm.

5. Pipe constant $= 1800$ cfm/83 fps $= 21.6$.

For 8-in. pipeline with an operating pressure of 6.6 psig is to be used.

For wheat, weighing 48 lb/cu ft, is to be conveyed over a distance of 400 ft at the rate of 20 tons/hr:

1. Use a saturation of 1.3, hp/ton of 2.6, and a pressure factor of 5.

2. Operating pressure $= \dfrac{2.6}{1.3} \times 5 = 10$ psig.

3. scfm $= 1.3 \times 667$ lb/min $= 870$ cfm.

4. acfm $= 870$ cfm $\times \dfrac{14.7}{(14.7 + 10)} = 517$ cfm.

5. Pipe constant $= 517$ cfm/55 fps $= 9.4$.

This pipe constant falls between 8.4 for a 5-in. pipe and 12 for a 6-in. pipe. While some lobe-type blowers have a maximum rating of 12 psig for continuous operation, most have a maximum rating of 10 psig. It is therefore inadvisable to increase operating pressure of the system to reduce the pipe constant down to that for a 5-in. pipe. We then use a 6-in. pipe and recalculate for the air and pressure requirements.

1. acfm = velocity times pipe constant:

$$55 \text{ fps} \times 12 = 660 \text{ cfm}$$

2. To find the absolute operating pressure for the 6-in. pipe, the following formula is used:

$$\text{psia} = \sqrt{\frac{\text{scfm(original system)} \times 14.7 \times \text{psia (original system)}}{\text{acfm (new system)}}}$$

$$= \sqrt{\frac{870 \text{ cfm} \times 14.7 \times (14.7 + 10)}{660 \text{ cfm}}}$$

$$= 23 \text{ psia,} \quad \text{or} \quad 8.3 \text{ psig}$$

3. $\text{scfm} = 660 \text{ cfm} \times \dfrac{(14.7 + 8.3)}{14.7} = 1070 \text{ cfm.}$

4. Saturation = 1070 cfm/667 lb/min = 1.55

With this new saturation factor of 1.55, and retaining the original hp/t factor of 2.6 and the pressure factor of 5, we can now check our calculations for the correct operating pressure:

$$\text{operating pressure} = \frac{2.6}{1.55} \times 5 = 8.4 \text{ psig}$$

This is within 1% of our previous calculation, so for all practical purposes, we can consider the operating pressure as correct. The 6-in. pipeline is to be used with the operating pressure of 8.3 psig. For actual blower requirements, the feeder leakage must be added to the scfm, calculated in step 3 above.

For soft feeds, weighing approximately 30 lb/cu ft, are to be conveyed over a distance of 400 ft at the rate of 50 tons/hr:

1. Use a saturation of 1.9, hp/ton of 3.7, and a pressure factor of 3.8.

2. Operating pressure $= \dfrac{3.7}{1.9} \times 3.8 = 7.4 \text{ psig.}$

3. scfm = 1.9 × 1667 lb/min = 3160 cfm.

4. acfm $= 3160 \text{ crm} \times \dfrac{14.7}{(14.7 + 7.4)} = 2100 \text{ cfm.}$

5. Pipe constant = 2100 cfm/70 fps = 30.

Since this is close to the constant for a 10-in. pipe (34), this size pipe should be used. We must now recalculate for the correct air requirements and operating pressure for this corrected pipe constant:

1. acfm = 70 fps × 34 = 2380 cfm.

2.
$$psia = \sqrt{\frac{scfm \times 14.7 \times psia}{acfm}}$$

$$= \sqrt{\frac{3160 \text{ cfm} \times 14.7 \times (14.7 + 7.4 \text{ psig})}{2380 \text{ cfm}}}$$

$$= 21.0 \text{ psia}$$

3.
$$psig = psia - 14.7$$

$$= 21.0 - 14.7 = 6.3 \text{ psig}$$

4. scfm $= 2380 \text{ cfm} \times \dfrac{(14.7 + 6.3)}{14.7} = 3400 \text{ cfm.}$

5. Saturation = 3400 cfm/1667 lb/min = 2.0

It appears the 10-in., schedule 30 pipeline should be used with an operating pressure of 6.3 psig. However, going back to the determination of the operating pressure from the factors now presented, that with a saturation of 2, the hp/ton factor remaining at 3.7, as well as the pressure factor at 3.8, we will get an operating pressure of 7 psig, which will be used for the moment.

We now wish to see what an 8-in. system will do since we have quite a difference between the manufacturer's maximum allowable blower pressure and the operating pressure of the 10-in. system.

1. acfm = 70 fps × 21.3 = 1490 cfm.

2. psia $= \sqrt{\dfrac{3160 \text{ cfm} \times 14.7 \times (14.7 + 7.4)}{1490 \text{ cfm}}} = 26.3 \text{ psia.}$

3. psig = 26.3 − 14.7 = 11.6 psig.

4. scfm $= 1490 \text{ cfm} \times \dfrac{(14.7 + 11.6)}{14.7} = 2665 \text{ cfm.}$

5. Saturation = 2665 cfm/1667 lb/min = 1.6.

6. Pressure check = 3.7/1.6 × 3.8 = 9 psig.

This pressure check can be most misleading when one is confronted with this large discrepancy between the 11.6 figure calculated earlier and the check figure of 9. It indicates that something more than formulas is required for the final design. The operating pressure of 11.6 psig is more nearly correct. In comparing this to the 7 psig pressure of the 10-in. system, it appears, further, that the experience factor, so needed with this type of conveyor, must be applied. This experience factor is knowing the feel of a conveyor under correct operable conditions. In this case, the author would increase the pressure on the 10-in. system to 7.5 psig and reduce the pressure on the 8-in. system to 11 psig. While formulas will work out about 80% of the time, the other 20%, when you rear back and take a second look, requires that knowledge gained only from experience, so difficult to put into words.

The principal difference between European and American design is the sizing of the conveying pipeline. American practice is to use the same size pipe throughout the system; European practice is to use a stepped pipeline (i.e., small at the entrance of material or pickup point, then larger at the terminal or discharge end). They use this method mostly on vacuum-type grain-handling systems.

This author's belief is that stepped pipelines are practical over long distances so that velocities are kept within reason and so that operating vacuums and pressures can be kept to a minimum. Also, the net result is lower power requirements.

How do we go about designing a long-distance pipeline? Without the benefit of many operating results, the following method has achieved good success: (1) determine first air and velocity requirements; (2) determine vacuum or pressure at 250-ft intervals along the pipeline; and (3) determine velocities at these intervals so that the step from one size to a larger size will maintain sufficient velocity to keep the material moving toward the discharge point.

For reasons known only to themselves, a mill desires to convey wheat discharging from a screw conveyor at 20 tons/hr a distance of 1500 ft using a vacuum-type system. The first assumption to make is the pipe size at the pickup point. For 20 tons/hr, an 8-in. schedule 30 seems reasonable. To accelerate the material in a short time, a pickup velocity of 115 ft/sec will be used.

The free air required is the velocity times the pipe constant:

$$115 \text{ ft/sec} \times 21.3 = 2450 \text{ cfm}$$

To see if we are in the ball park, let us check our saturation figure:

$$\frac{2450 \text{ cfm}}{667 \text{ lb/min}} = 3.67$$

This seems reasonable for a distance of 1500 ft, since the saturation table (Table 3.6) with correction factors for 1000-ft conveying distance gives us a figure of 3.10.

Using the following formulas for flow of air in pipes, we can closely determine the light load (air only) vacuum at various points along the pipeline.

$$v = \sqrt{\frac{25,000dp}{l}}$$

$$p = \frac{lv^2}{25,000d} \times 0.65$$

$$V = vA$$

where v = velocity of air (ft/sec)
        p = loss of pressure due to flow (oz/sq in.)
        d = inside diameter of pipe (in.)
        l = length of pipe (ft)
        V = volume of air (cf/sec)
        A = area of pipe (sq ft)

The correction factor of 0.65 in the formula for pressure loss is a correction factor that takes into account the less dense air in the pipe (below 14.7 psia). While this correction factor can vary for each foot of pipeline length, experience has shown it to be applicable for the complete pipeline system.

For pressure loss when conveying material, we can use the following empirical formula which is quite accurate where A over M lies between 3 and 10 and W lies between 1 and 2.

$$p_1 = p_2 \left(1 + \frac{(1 + 0.000285L)\ W^2}{\sqrt[3]{\dfrac{A}{M}}} \right)$$

where p_1 = pressure loss loaded
        p_2 = pressure loss light
        A = cu ft of air per minute (cfm)
        M = pounds of material per minute
        W = specific gravity of material (sp gr)
        L = length of conveying line in feet

Referring to The Consolidated Grain Milling Catalogs (6th ed., 1945–1947) we find that wheat has a specific gravity of 1.355–1.423, and in bulk it has a specific gravity of 0.720–0.840. For practical purposes, let us use a figure of 1.0, the average of the two lowest values.

From these two formulas we calculate the light and loaded vacuums, the free air at the points considered and the velocity within the 8-in. pipe and one size larger, 10-in. Also, let us presume a pickup loss for acceleration of the material within the pipeline of 4 oz/sq in.

The results are tabulated as follows:

Point in Conveying Line		Vacuum (psig)			Velocity (fps)	
		Light	Loaded	cfm	8-in.	10.in.
8-in.	Intake	0.00	0.25	2492	117	
	250 ft	0.69	1.17	2662	125	78
	500 ft	1.50	2.53	2959	139	87
	750 ft	2.47	4.13	3407	160	100
10-in.	750 ft	1.81	3.04	3094		91
	1000 ft	2.14	3.58	3239		95
	1250 ft	2.50	4.17	3420		101
	1500 ft	2.91	4.84	3653		107
Static loss filter-receiver		1.00	1.00	4065		
total opera-ting vacuum		3.91	5.84			

(7.82 in. Hg) light; (11.68 in. Hg) loaded

Assume the system to operate at 12 in. Hg.

From the preceding tabulation, it appears feasible to enlarge the pipeline from 8 in. to 10 in. at or a little downstream from the 500-ft point using a truncated cone shape having a minimum length of 10 ft. This should either eliminate or minimize the turbulence in going from one pipe size to the next larger size. It is very improbable that the wheat itself has reached the velocity of the air at the 500-ft point. It should, however, have sufficient velocity to remain in suspension in the airstream when reaching the 10-in. pipe.

For blower size:

$$\frac{4065 \text{ cfm}}{0.85} = 4782 \text{ cfm}$$

Referring to catalog data, we find a size 1024 having a full rated pressure of 7.9 psig, a maximum cfm of 5332 at 1770 rpm and 6.0 psig, and a displacement of 3.3 cf/r.

Blower speed:

$$1770 \text{ rpm} - \frac{5332 \text{ cfm} - 4782}{3.3 \text{ cf/r}} = 1603 \text{ rpm}$$

Slippage:

$$3.3 \text{ cf/r} \times 1770 \text{ rpm} = 5841 \text{ cfm}$$

$$\frac{5841 \text{ cfm} - 5332}{3.3} = 154 \text{ rpm}$$

Slippage due to vacuum:

$$(154 \text{ rpm} \times 1.241) - 154 = 37 \text{ rpm}$$

Total rpm: 1603 + 37 = 1640
Blower hp:

$$1640 \text{ rpm} \times 3.3 \text{ cf/r} \times 6.0 \text{ psig} \times 0.005 = 162 \text{ bhp}$$

These results compare favorably with known installations.

Let use see what will happen if we can use a low-pressure-type system (i.e., up to 15 psig). First, we notice that the velocity spread between the pick-up and delivery points is much less than that for the vacuum-type system. Using a velocity of 55 fps at the intake, the velocity at the discharge is as follows:

$$\frac{55 \times (14.7 + 15.0)}{14.7} = 111 \text{ fps}$$

If we assume a 6-in. pipeline, the acfm required at the pick-up point will be 55 fps × 12, or 660 cfm. If we assume an 8-in. pipeline, the acfm required is 55 × 21.3, or 1171 cfm. For the same conveying rate of 20 tph, we have a rate of 667 lb/min. The A (cfm) over M (lb/min) in the conveying pressure loss formula in this case varies from 1.0 to 1.75. This is below the range of 3 and 10 at which the formula is effective. A stepped pipeline in either 6-in. or 8-in. pipe is not necessary. As a precaution against kernel breakage, the final 400 ft might be enlarged. It appears then that low-pressure systems require no stepping at all.

In medium- and high-pressure systems (15–45 and 45–90 psig operating pressures), the pipes are sometimes stepped when the conveying distance exceeds 1600–1700 ft to maintain a terminal velocity of from 66 to 100 ft/sec. With low velocities prevalent in these systems, stepping is not all that necessary. Sometimes it is only a personal inclination.

3.3 SYSTEM COMPONENTS

In choosing the various components that enter into the complete pneumatic conveyor system, the following descriptions are given to indicate the different types available. Although some may be superior to others under similar operating conditions, no effort is made to discuss their relative merits. Each manufacturer makes claims for his equipment that for the most part are based on his experience in the handling of the same or similar materials under comparable operating conditions. Substantiation of such claims is usually offered by referring to previously installed, successfully operating systems.

Blowers

The most-used blower in pneumatic conveying service is the rotary positive type having two figure-eight impellers rotating in opposite directions within a closely machined housing (Figure 3.1). As each impeller passes the blower inlet, it traps a definite volume of air and carries it around the case to the blower outlet, where the air is discharged. This type of blower has a positive displacement plus a constant volume, operating at varying pressures. With each revolution, the blower delivers a metered amount of air measured at inlet conditions. Operating at a constant speed against a constant pressure, the blower delivers a constant amount of air. Increasing the blower speed against a constant pressure increases the volume by an amount equal to the displacement of the blower times the increased number of revolutions.

Historicially, the two-impeller, or lobe-type positive blower was developed somewhat by accident. Around the middle of the nineteenth century, the brothers P. H. and F. M. Roots, who owned a textile mill in Connersville, Indiana, needed a water turbine to drive the lineshafts by the fall of water from the canal adjacent to the mill. A satisfactory water turbine had not yet been built in those days, so F. M. Roots designed a two-impeller contraption with a sheet metal case and wooden impellers. His idea was fundamentally sound, but his choice of materials of construction doomed his machine to failure. The failure was caused by the wooden impellers swelling from the water, jamming the turbine.

After considerable scraping to the wooden impellers, the turbine was belted to the existing lineshaft and turned over to dry out. The local bewhiskered foundryman, curious about this new machine, looked into the

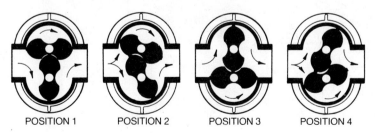

POSITION 1   POSITION 2   POSITION 3   POSITION 4

FIGURE 3.1.   Rotary positive blower operating principle. (Courtesy Roots Operations, Dresser Industries, Inc., Connersville, Indiana.)

top—and his hat blow off. "This will make a better blower than it will a turbine," he said. Thus was born the first Roots blower in 1859.

Basic patents were granted in the name of P. H. and F. M. Roots, who began the manufacture of such blowers and who also licensed European manufacturers to build Roots blowers. The two-impeller type, positive-pressure blower is still called the Roots blower in Europe. In America, only blowers made by the original builders and their successors are called Roots. It is interesting to note that the rotary positive blower as it is manufactured today is the Roots blower in principle, plus refinements in design, and with all the improvements in materials and workmanship permitted by today's technological developments.

Blowers today are manufactured in many sizes and types, and in varying degrees of refinement and allowable speeds. Manufacturer's catalogs should be consulted and their recommendations followed. For general conveying applications, blowers with minimum or no shaft seals in the headplates can be used, but handling inert gases, where losses at this point can be critical, blowers with heavy-duty shaft seals should be used.

Positive-pressure rotary blowers will deliver varying amounts of air from a few cubic feet to as much as 32,000 cfm, and at pressure normally up to 10 psig. Some sizes will go as high as 15 psig, but these are somewhat limited.

Another type of blower used for pressure, vacuum, or combination service is the axial flow, positive displacement blower/compressor (Figure 3.2). Like the lobe-type blower, there is no need for lubrication within the compression chamber, again assuring clean air or gas delivery. The axial flow blower uses a screw-type cycloidal rotor, which discharges its air in a smooth and steady fashion. Its range of operation is from a few cubic feet of air to approximately 12,000 cfm, and at pressures up to 18 psig.

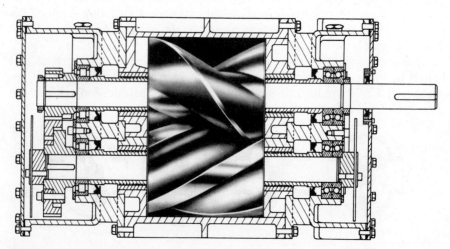

FIGURE 3.2.   Axial flow positive-displacement blower/compressor. (Courtesy Gardner-Denver Co., Quincy, Illinois.)

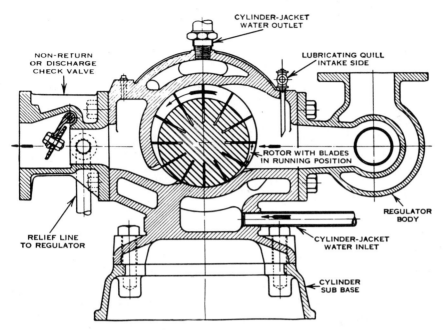

FIGURE 3.3. Rotary multivane-type compressor. (Courtesy Fuller Company, Bethlehem, Pennsylvania.)

For pressure applications above 15 psig, the rotary, multivane-type compressor (Figure 3.3) is ideally suited to the activation of the medium- and high-pressure systems. This type of compressor is manufactured in both single- and two-stage construction, having a range in capacity from 30 to 3300 cfm, and at pressures up to 125 psig. Because of its rotary design, this type of compressor delivers a steady, nonpulsating stream of air directly from its discharge port into either the manifold of a fluid solids pump or a blow tank, without the need for an intermediate air receiver. Maximum economy of air compression is achieved since any line pressure variation in the conveying system is reflected back to the air compressor, causing the unit to compress to the exact air pressure requirement.

The centrifugal turbocompressor, in multistages (Figure 3.4), is used successfully on both vacuum- and pressure-type conveying systems. Maximum capacities of such systems is about 12 tons/hr with the pipeline sizes one size larger for the same conveying rate than when using a rotary positive blower. Power requirements, however, will be about the same. Lower pressures (psig) and vacuums (in. Hg) are traded off for higher air volumes (cfm).

One of the centrifugal compressor's greatest advantages is its quiet operation. This is due in part to the multistage construction with the accompanying low peripheral speeds that in practically all cases are kept to a minimum. Air is delivered with a uniform pressure but in varying volumes, depending on the static load from the conveying system. A blast gate at its

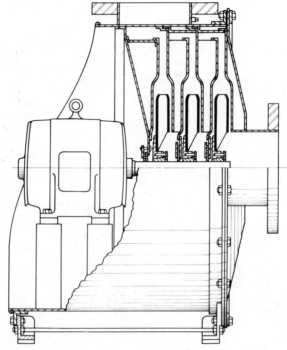

FIGURE 3.4.   Centrifugal turbocompressor. (Courtesy The Spencer Turbine Company, Windsor, Connecticut.)

discharge is usually sufficient to maintain air volume necessary to maintain a conveying condition.

Industrial-type and material-handling fans are used for venturi-type low-pressure systems and waste-handling systems. The industrial type of fan has a capacity of up to 12,000 cfm at 30 in. static pressure; the material-handling fan has a capacity of up to 20,000 cfm at 18 in. static pressure. Both types use impellers specially designed for passing materials through the fans.

One other air mover must be mentioned, not because of its universal use but because of its sporadic use for large pressure systems, notably wood chip systems. It is the single-stage centrifugal compressor for air volumes from 10,000 to 30,000 cfm (Figure 3.5). While the rotary positive-type blower is a constant-volume, varying-pressure machine, the centrifugal compressor is a variable-volume, constant-pressure unit. To maintain a relatively constant volume and to control surges, a partially closed intake controlled through instrumentation is necessary to maintain airflow through the pipeline system.

Study of a few existing systems indicates that the designers, after determining the pressure required for the conveying system, chose a centrifugal blower to operate at about 10% higher pressure. This, evidently, is to assure sufficient pressure to maintain a conveying condition within the pipeline and to avoid a plugged system. While a centrifugal compressor is less costly than a comparable rotary positive blower in volume and pressure delivery, experience has demonstrated that the rotary positive blower is more reliable when coping with the variations in line pressure in a pneumatic conveying system, while maintaining a constant volume.

To comply with OSHA (Occupational Safety and Health Administration) regulations and to maintain good community relations, rotary positive blowers in pneumatic conveying service should be silenced so that the sound level (dBA) of their intake and discharge does not exceed 90. For pressure systems this may require both inlet and discharge silencers, in addition, it may require lagging of the piping in the vicinity of the blower to reduce noise radiation. Vacuum systems normally require only a discharge silencer. Silencer manufacturers such as Burgess Industries and Universal Silencer offer excellent design and engineering information.

Regarding community relations, some years ago a pulp mill in northern New England installed a wood chip system to convey chips from the screens in the woodroom to the chip loft above the digesters. The pressure-type system terminated in a cyclone receiver with just a weatherhood over its top air discharge outlet. The blower was equipped with only an intake screen and silencer. This took care of the blower, since the noise from the wood-

FIGURE 3.5. Single-stage centrifugal blower. (Courtesy Allis-Chalmers Corporation, Milwaukee, Wisconsin.)

room was much louder than that generated by the blower. The blower discharge pulsations, however, were not muffled very much by the conveying pipeline and the chips being conveyed through it. Noise from the cyclone discharge was most annoying, especially when the wind was in the right direction. The noise traveled about a half mile into town, directly to the local hospital. Complaints were many, right up to the town council—and very uncomplimentary. The addition of a discharge silencer corrected the problem.

The centrifugal turbocompressor does not require silencers, nor do fans, except in isolated cases, where noise levels are beyond acceptable limits, such as in a confined working environment. To reduce the noise level, suppression devices are installed on the discharge side of the fan. The single-stage centrifugal compressor should be equipped with an intake silencer. Its intake noise can be likened to a turboprop airplane. Discharge silencers are normally not needed.

Filter-Receivers

Filter-receivers have been used for separating the material from the airstream in vacuum systems for about 60 years. In going through the archives, the earliest use was indicated in 1922. The filter media consisted of mostly cotton and muslin sheeting. Originally, they were of the single-compartment type; next came the two-compartment, followed by the four-compartment type. The filter media, like those used today, consisted of cylindrical tubes. Cleaning of the tubes was done by reversing air through them.

In the development of the number of compartments in the filter sections, it was found in a single-compartment filter that in order to clean the bags, the conveying air had to be shut off, and that shaking only was practical for the cleaning action. With the advent of the two-compartment unit, it was possible to maintain conveying by cutting one compartment out of service for cleaning while the other compartment remained in service. The compartment was taken out of service for a period of 15–30 sec, with the cleaning cycle taking place every 2 or 3 min. It was found, however, that the actual amount of cloth used to separate the material from the airstream, and to retain dust from discharging to atmosphere during cleaning operations, allowed a very high amount of air to pass through the smaller area of filter media, causing, at times, severe bleeding. This difficulty led to the design of the four-compartment filter in the late 1920s.

The four-compartment filter section provided a higher ratio of active filtering area during the cleaning cycle and still retains high popularity. In addition to air reversal, shaking of the filter bags was done during the same time, to give additional cleaning efficiency. Actuators for the bag cleaning shaking mechanisms include airmotor, hydraulic operators, and fractional horsepower electric gearmotors. Through limit switch interlocking and positive cam type timer operation, these units became self-contained and automatic in operation. Because of the shaking of the filter bags, the filter media is restricted to either cotton, wool, or such synthetics as Nylon, Orlon, and Dacron, all of the woven type.

The four-compartment filter (Figure 3.6) indicates the general arrangement and design features. For dusty nonabrasive materials as well as grains, the tangential inlet is used to give a simulated cyclonic separation in the bottom cone. For abrasive materials, to resist wear to the exterior shell, radial intakes are used, with some type of wear-resisting back for the material to hit against, or a pocket is constructed so that the material will impinge on itself (Figure 3.7).

To facilitate a cyclonic action in the bottom cone, an inner skirt is provided to create this condition as well as to eliminate the possibility of the conveying air being short-circuited directly into the bottom crown sheet and into the filter bags. Likewise, in the four-compartment unit, the partition plates dividing the unit into these four compartments should extend down to the bottom of the inner skirt. For proper operation and minimum short circuiting between the intake and the filter bags, the inner skirt should extend 4–6 in. below the lowest part of the intake opening.

The crown sheet to which the open bottom of the filter tubes is attached should be a fairly thick steel plate, with the necks securely welded and airtight. With the number of holes cut into this plate to accommodate as many filter tubes as possible within a given circle, the plate very much resembles a piece of Swiss cheese. In this condition, its structural strength is considerably weakened, and much flexing will result during operation. By welding the necks into the crown sheet, much of the original strength is regained, with flexing in operation reduced to the minimum. It must be remembered that when the system is in operation and being operated at a vacuum of 12 in. Hg, the differential at times across the one compartment that is out of service for cleaning is 6 psi, or, to be more impressive, 864 psf.

Since the shaking mechanisms impart a somewhat violent shaking reaction to the cloth tubes, most of the wear on the tubes occurs in the bottom third. At this point, additional reinforcing may be necessary, either through a multiwall bag or additional spreader rings. With the multiwall bag, however, this space between the walls may fill up with material or dust, causing an additional load to be placed on the shaking mechanism, obviating any gain from the multiwall principle.

Such filters, are usually sized by the square feet of cloth required to retain the dust for the cubic feet of air being passed through it. Depending on the material being conveyed, together with its dust loading, this figure may vary from 1 up to 6. It is also necessary to minimize the velocity of the air as it passes up through the necks of the crown sheet to minimize the blasting effect of the bottom part of the filter tubes. At times, this blasting effect can be more detrimental than the flexing resulting from the action of the shaking mechanism. With the advent of felted fabrics, it is entirely within

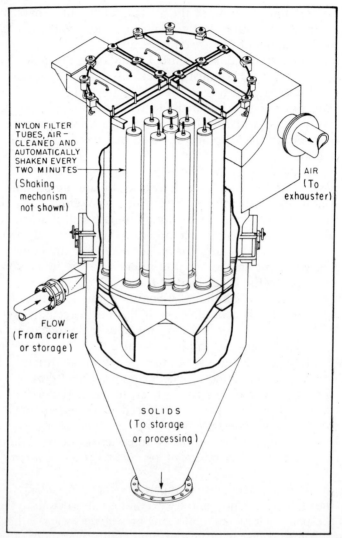

NYLON FILTER
TUBES, AIR-
CLEANED AND
AUTOMATICALLY
SHAKEN EVERY
TWO MINUTES
(Shaking
mechanism
not shown)

AIR
(To
exhauster)

FLOW
(From carrier
or storage)

SOLIDS
(To storage
or processing)

FIGURE 3.6.   Four-compartment filter-receiver. (Courtesy, Fuller Company, Bethlehem, Pennsylvania.)

the realm of possibility that this type of filter, using air reversal only for cleaning the cloth tubes, can be a very efficient separator and that there can be a reduction in its maintenance cost. The air-to-cloth ratio, however, would probably fall between those listed in Table 3.11 for the woven and felted fabrics.

With the development of the Mikro-Pulsaire* collector using felted filter fabric about 25 years ago, a new dimension in filter-receivers was brought

* Registered trademark, Mikropul Corp., Division of U.S. Filter Corp., Summit, New Jersey.

on the market (Figure 3.8). This type filter-receiver collector consists of the usual inlet, bottom cone, and inner skirt. The material being conveyed enters the bottom cone through the inlet, with the heavier particles discharging out through the bottom outlet. The dust in the conveying air travels upward uniformly with the air, about the filter bags. The dust is retained on the outside surfaces of the filter bags, and as this dust material builds up on the bags, periodic cleaning becomes necessary. Cleaning is accomplished in this unit by the process of aerodynamic valving. This valving, controlled by an adjustable sequence timer, allows solenoid-operated valves to progressively cycle momentary bursts of compressed air at 100 psig and go through an internal piping system in the clean air plenum. Through a series of precisely located orifices directly above the individual filter bags, the compressed air is directed down into the filter bags, setting up a shock wave that flexes the filter cloth and dislodges the dust cake on the exterior of the bags. The dust cake drops into the bottom cone, is mixed with the heavier material, and is discharged with it through the bottom outlet.

In cold climates, dry air should be used in this type filter-receiver. If freezing temperatures occur, and normal plant air containing moisture is used without removing this moisture, freezing can occur in the various valves, which could cause some operating problems. In normal temperatures, where no freezing occurs, the probability of having problems from this source is practically nil.

Another filter-receiver (Figure 3.9), also using felted fabric filter tubes,

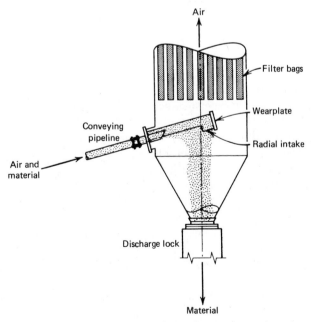

FIGURE 3.7.   Radial intake for abrasive materials.

TABLE 3.11.   Air-to-Cloth Ratios, Filter-Receivers

	Filter Media	
Material Conveyed	Woven	Felted
Alum	5–1	8–1
Alumina	3–1	5–1
Carbonate, calcium	4–1	6–1
Cellulose acetate	5–1	8–1
Clay, air floated	4–1	6–1
Clay, spray dried	4–1	6–1
Clay, water washed	4–1	6–1
Coffee beans	5.5–1	10–1
Corn, shelled	6–1	10–1
Flour, wheat	4.5–1	9–1
Grits, corn	5–1	8–1
Lime, pebble	5–1	8–1
Lime, hydrated	4–1	6–1
Malt	6–1	10–1
Oats	5.5–1	9–1
Phosphate, trisodium	3–1	5–1
Polyethylene pellets	6–1	10–1
Rubber pellets	3–1	5–1
Salt cake	3–1	5–1
Soda ash, light	4–1	6–1
Soft feeds	5–1	8–1
Starch, pulverized	4–1	6–1
Sugar granulated	4–1	6–1
Wheat	6–1	10–1
Wood flour	4–1	6–1

has its own built-in reverse air supply—no outside source of compressed air is required. This air reverse uses a small centrifugal blower (2–7.5 hp) to produce a high-volume, low-pressure airstream to a rotating arm in the top section of the filter-receiver. This rotating arm, driven by a $\frac{1}{3}$-hp gear-motor and drive mechanism, operates at a speed of one revolution per minute so that each filter tube is pulsated by a burst of high-velocity air each minute to dislodge the material and dust collected on the outside of the filter tube. It too uses an inner skirt below the filter tubes to give cyclonic action to the incoming material and air in the bottom cone and to prevent the incoming airstream from short-circuiting up into the filter compartment. Unlike the four-compartment filter-receiver, but like the Mikro-Pulsaire filter-receiver, no partition plates are required, since the filter compartment is a single unit.

Whatever choice of filter media is made, it is wise to check what the manufacturer's recommendation would be for the proper air-to-cloth ratio for the material being conveyed. As a guide, Table 3.11 indicates the max-

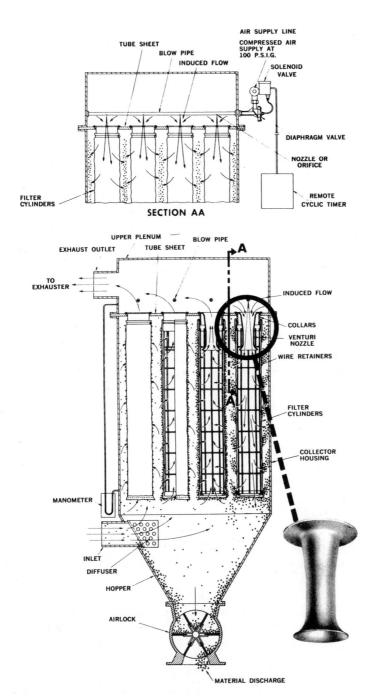

FIGURE 3.8. Mikro-Pulsaire filter-receiver. (Courtesy Mikropul Corporation, Division of U.S. Filter Corporation, Summit, New Jersey.)

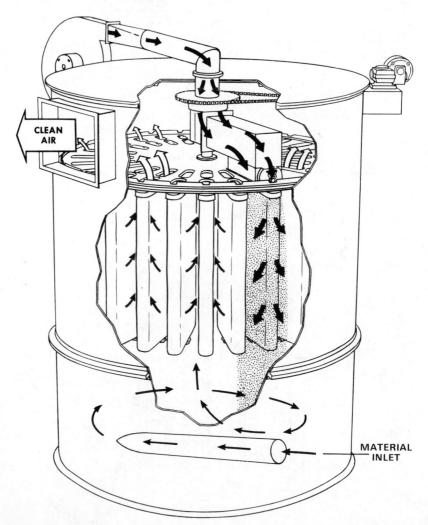

FIGURE 3.9.   Reverse-air filter-receiver. (Courtesy CEA—Carter-Day Company, Minneapolis, Minnesota.)

imum air-to-cloth ratios for the two types of filter media and filters described heretofore. Cloth area is the net or smallest area in service during the time the bags are being cleaned. The air volume used should be the scfm as determined in the vacuum-type system.

In Section 3.1, in the criteria required for design and use, two characteristics of the material being conveyed were to be determined. These were the pH and the temperature. These two characteristics determine the type of filter cloth to use. From the values received and tabulated, the proper cloth can be determined from the comparison chart (Table 3.12).

In some cases it is desirable to separate as much dust as possible from

TABLE 3.12.   Filter Cloth Comparison Chart

Filter Cloth	Temperature Limit (°F)	pH of Material Conveyed	
		2.0–6.5	5.5–11.0
Cotton	180°		x
Dacron[a]	275°	x	x
Dynel[b]	180°		x
Nomex[a]	375°		x
Nylon	230°		x
Orlon[a]	275°	x	
Polypropylene	150°	x	x
Teflon[a]	400°	x	x
Wool	210°	x	

[a] Registered trademark, E. I. duPont de Nemours & Company, Inc., Wilmington, Delaware.
[b] Registered trademark, Union Carbide Corporation, Danbury, Connecticut.

the granular materials being conveyed before they are deposited in storage or process without further cleaning. Sometimes a separate cyclone is used to separate the granular material from the airstream, with the dust-laden air terminating in a filter-receiver. Where space is at a premium, an adaptation, combining the primary cyclone and the filter-receiver, can be made through the addition of a secondary cyclone in the bottom cone of the filter-receiver to collect the dust as it is dislodged from the filter tubes. This dust is separately discharged through its own outlet (Figure 3.10). While

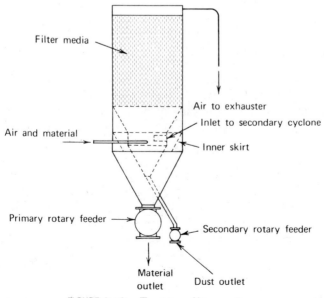

FIGURE 3.10.   Two-stage filter-receiver.

this increases the height required for the filter-receiver, it does reduce the required floor area that is necessary for both the separate cyclone and the filter-receiver.

In operation, the conveying stream enters the primary cyclonic bottom cone with the granular material discharging out through the bottom cone outlet, and the dust-laden air then is brought into the internal secondary cyclone where some additional cyclonic separation is accomplished with the dust-laden air going up into the filter bags. As the dust is dislodged from the filter bags, it drops down into the internal cyclone and is discharged out through its own outlet. The dust now is completely separated from the mainstream of material. This arrangement is used principally in receiving stations for systems handling grain, where it reduces the separation requirements of the grain cleaners and allied equipment necessary to remove the dust that is detrimental to the process that follows. Care must be exercised, however, in the size of the downspout from the internal cyclone as well as the size of the rotary feeder or discharge lock at its outlet. They must be large enough so that sufficient area is available so that the blowback air through the discharge lock can dissipate up through the spout without impeding the flow of dust down to the rotary feeder or discharge lock.

When driving the rotary feeders on both discharge outlets, either a single-chain drive should be used to operate both feeders from a single driver, or, if separate chain drives are used, the second feeder should be driven from the tail shaft of the first. Operation of these feeders is a prime requirement for the elimination of backup of material in the cyclones and filter. A motion switch, driven from the tail shaft of the last driven feeder in the train, should be used and interlocked with the exhauster motor to prevent it from activating the system when the feeders or discharge locks are inoperative. Similarly, if only a single feeder or discharge lock is used below the filter-receiver, it too should have a motion switch driven from the tailshaft of the feeder.

Filter-receivers, properly designed and sized with a conservative air-to-cloth ratio, can be the best deterrent to air pollution not only at their point of use, but also in the surrounding area. You cannot be too conservative in the amount of filter cloth used, nor in eliminating the maintenance required by these units.

Cyclone Receivers

Cyclone receivers are used for air–material separation at the terminal end of both vacuum and low- and medium-pressure systems. They are not 100% dust retentive; therefore their use is limited to materials of a nondusty nature. Contrary to this, however, they are used with quite a high degree of success at the terminal end of medium-pressure systems where the velocity is nowhere near as great as that encountered in vacuum and low-pressure systems. Their use at this point in pressure systems is enhanced by their increased efficiency caused by not having the cyclonic action interrupted by the blowback of air through the bottom outlet and discharge lock, so prevalent in the vacuum system.

Cyclone receivers for use in pneumatic conveying are designed and built in two types, single-stage (Figure 3.11a) and double or two-stage (Figure 3.11b). Sizing of cyclone receivers, like power and air figures, has been developed for the most part from experience, with much theory and practice added to the knowledge gained from that experience. In dust collecting especially, the added theoretical designs supplemented the designs gained from the early experience. Scroll-type inlets and air outlets enhance the efficiency of dust-collecting cyclones, where this efficiency is deemed so necessary. In pneumatic conveying, the plain designs have sufficed to give

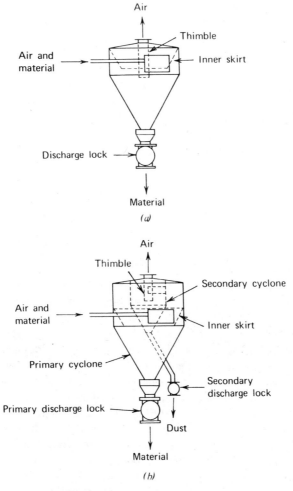

FIGURE 3.11. Cyclone receivers.

TABLE 3.13.   Cyclone Receiver Selection

| | Diameter Cyclone Receiver | |
Conveying Line Diameter	Primary	Secondary
3 in.	3 ft, 6 in.	2 ft, 9 in.
4 in.	4 ft, 0 in.	3 ft, 0 in.
5 in.	4 ft, 6 in.	3 ft, 6 in.
6 in.	5 ft, 0 in.	3 ft, 9 in.
8 in.	6 ft, 9 in.	5 ft, 0 in.

a satisfactory efficiency percentage with the bottom cone angles varying from 60 to 75° for primary cyclones and from 70 to 85° for secondary cyclones. Typical cyclone sizes (diameters) for conveying pipeline diameters are indicated in the cyclone receiver selection table (Table 3.13). Primary receivers are for vacuum systems only, but secondary receivers are for both vacuum and pressure.

Materials of construction and sturdiness of construction are determined by the material being conveyed and its effect on the wearability of the metal and the thickness chosen. In handling abrasive materials, wear-resisting liners are added at times to extend the useful life of the cyclone receiver, especially at points of impact.

Rotary Feeders

Rotary feeders for pneumatic conveying service should always be of heavy-duty construction. Many types are available. Whether it is to be used for vacuum or for pressure service, the feeder is a very vital part of the whole system. Therefore both the designer and the user should carefully choose the correct type.

What should we look for in a rotary feeder? Before considering this, we must know the material to be passed through the feeder, especially its physical state. Then we must know the volume of material, the differential across it, and the service it is to perform.

Knowledge of the material and its physical state is necessary to determine the size of the feeder and its speed (revolutions per minute). Frequently the bulk density (pounds per cubic foot) is known or, if not known, can be easily determined by simple measurement in its normal state. But it is very unlikely that this is the state at which the material will enter the feeder. Materials that are aeratable or that can be fluidized will have a much lower bulk density in the aerated state, hence much higher volume for the same total amount or weight. A ton of material having a bulk density of 50 lb/cu ft will have a volumetric displacement of 40 cu ft. When this same material is aerated and has a bulk density of 40 lb/cu ft, its volumetric displacement is increased to 50 cu ft. To size the feeder correctly, the bulk density as it enters the feeder must be known.

A good example of this change in bulk density was presented by the unloading of spray-dried clay from a covered hopper car and its delivery to a mixing tank. The clay passed through a vacuum-type system using a rotary feeder below the filter-receiver, and an automatic scale, fed by a screw feeder, to meter the clay into the mixing tank. A sample of the clay taken from the top of the car indicated that at rest the clay had a bulk density of 49.3 lb/cu ft. After traveling through the 100-ft conveying pipeline, the filter-receiver, and the rotary feeder discharge lock, the clay was extremely fluid from aeration. As it discharged from the lock, its bulk density was reduced to 23.7 lb/cu ft. Fortunately, the feeder was sized to handle the aerated material. If it had instead been sized on the at-rest density of the clay, it would have been too small, necessitating replacement, at considerable cost, plus loss of production, and a very much dissatisfied operating department. Even a 100% factor of safety would have been inadequate.

The differential across the feeder is necessary so that the shaft and rotor are of such design that the deflection due to this differential will not bind the rotor within the closely machined housing. This differential in the case of both vacuum and pressure systems was determined in the design of the system. In the vacuum system, the vacuum at which the system will operate is the differential across the feeder or discharge lock below the filter-receiver, since we normally have atmospheric conditions outside the filter-receiver and the outlet flange of the discharge lock, and the vacuum within the filter-receiver. In the pressure system, the pressure at which the system operates is in the pipeline, below the rotary feeder, exerting a pressure on the rotor since the inlet flange of the feeder is normally at atmospheric conditions.

Rotary feeders, having complete metal rotors and bodies and machined to very close clearances between the outside diameter of the rotor and the inside diameter of the body, require that the deflection of the shaft and rotor be extremely small. Most feeders are finished to a clearance of about 0.006 in. on the periphery and slightly less between the rotor and the head-plate. This is the star wheel rotor (Figure 3.12), which is the most prevalent used. Along with shaft diameter, the packing glands and location and size of the bearings are very much governed by this pressure differential.

Many variations in the design of star wheel rotors are available. Since this part of the rotary feeder is really its basic part, its design will determine the adequacy of its performance. For maximum capacity, the solid vane rotor with deep pockets should be used. The number of vanes is determined by the amount of sealing required between the rotor and the body. The minimum number of vanes needed to have two vanes seal between the inlet and the outlet of the body of the feeder is normally six. When more vanes are added a problem arises for with each additional vane, the angle between the vanes becomes smaller. This is reflected in a narrower throat where the vanes meet toward the longitudinal centerline of the rotor. The narrower

1. Six vanes. Two vanes seal between inlet and outlet.

2. Tapered rotor hub. Reduces friction from trapped material between rotor and headplates.

3. Integral-cast bearing brackets insure concentric bearing centers with male-female joint between headplate and feeder body.

4. Bearing brackets on sides of packing glands allow free drop of packing-gland leakage; prevents material from entering bearings.

5. Integral-shielded ball bearings; factory greased for life of bearing.

6. Shouldered shaft for externally centering and maintaining alignment of rotor, resulting in equal rotor end clearances and eliminating contact between rotor and headplates.

7. Heavy body walls and headplates. Rigid construction to permit smallest possible rotor clearances. Permits at least one possible remachining after normal wear.

8. Standard companion flange size and bolt circle permits connection to pipe size flanges.

9. Integral cast lugs for drive provides universal mounting for any type motor reducer. Rigid drive for minimum space requirements. Eliminates drive stress between feeder and related equipment.

10. Shear pin coupling with simple standard bolt shear pin.

11. Gear motor with right angle combination drive. Can be furnished with a parallel shaft drive if desired.

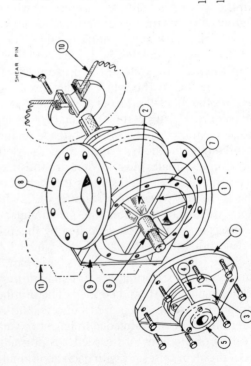

FIGURE 3.12. Drop-through rotary feeder with star wheel rotor. (Courtesy Fuller Company, Bethlehem, Pennsylvania.)

92

this throat becomes, the more difficult it is for material to drop out of that part of the rotor. A larger fillet, or heavier hub, is necessary to avoid this condition. This, however, will reduce the volumetric capacity of the rotor. If we continue this pattern, we will arrive at the shallow pocket type of rotor. While giving a cleaner discharge, it will greatly reduce capacity.

If we find the material to be sticky, or even balky at dropping out of the rotor, the surfaces can be sprayed with liquid Teflon and then baked at a high temperature. The resulting finish is smooth, but soapy in texture, like that found in Teflon-lined cookware, and it will not crack, chip, or peel. The low coefficient of friction of this Teflon coating will allow almost all sticky materials to drop out freely.

Where mildly abrasive materials are encountered, rotors with Stellited tips can be provided to reduce wear to the rotor. This does not mean that wear to the body will be reduced. In such instances, it is desirable to have the body cast of the highest abrasion resisting metal that can still be machined to close tolerances.

Although it is not entirely applicable to mildly abrasive materials, the addition of separate tips to the rotor blades can reduce wear and the high maintenance encountered in many cases. Rotor tips of many materials can be used. The most prevalent are neoprene, nylon, Teflon, and quite often brass. These tips are fastened to the leading edges of the rotor blades, and for the most part are adjustable. The thickness of these blades is usually no more than $\frac{1}{4}$ in. For some materials being handled by the rotary feeder, this should be reduced to not less than $\frac{1}{8}$ in., the minimum for structural strength.

In the solid rotor, the separate tips can be simulated to a great extent by the beveling of the trailing edges of the blades. In a way, this is superior to the separate tips when handling smudging and sticky materials, since the beveling is done not only on the periphery of the blades but also on the ends. Moreover, beveling does not affect the close operating clearances between the rotor and the body. It does allow the rotor to operate at a greatly reduced frictional drag, which is reflected in lower power requirements. It also promotes a self-cleaning operation since the rotor acts as a scraper, reducing the amount of buildup on the stationary parts (body and headplates) in which the rotor operates. The land, or peripheral edge of the rotor, to maintain sealing tolerances, should be $\frac{1}{8}$ in. wide. The bevel should have a minimum of 10°, with the maximum angle about 30°.

When special materials of construction such as stainless steel, brass or bronze, monel, or other corrosion-resistant materials are used, it is desirable to have all wear occur to the rotor. In rebuilding and maintaining a rotary feeder, it is much easier to retain original clearances through the rotor rebuild than through the body and headplates. If wear of any type occurs to the body and headplates, it is nearly impossible to build up the metal and remachine. All we can do is machine out the imperfections. When this becomes necessary, we must have a feeder whose parts have sufficient thickness to allow us to do the necessary machining and still have enough metal left to give the feeder body the structural stability needed to remain round, and to retain its torsional resistance. In fact, whether or not this remachining will be necessary, when purchasing the feeder it is wise to determine whether or not its construction is heavy enough to stand what you expect it to do.

To avoid this machining out the body and resurfacing the headplates, it is necessary to make the rotor expendable. This can be done not only by the addition of separate tips, but also by building up the rotor blade tips with a softer metal than that used in the body and the headplates. The softer material will wear, and when clearances are widened it can be rebuilt through welding or brazing. After the buildup of material is completed, it can then be turned down to the proper diameter and tolerance in a lathe. Similarly, if the body needs truing, the amount taken off to do this can simply be added to the rotor diameter to maintain the proper clearances and tolerance. In addition to the ease of rebuilding, this building up the tips of the rotor with material will give us the advantage of beveled rotor blades. The buildup of material is usually no greater than $\frac{3}{16}$ in. wide, which leaves the rest of the blade below the outer periphery.

Unlike the rotor, the construction of the body and the headplates is dictated by more than the material being conveyed. In addition to this, the purpose for which the feeder will be used will determine the body design and type. Types include drop-through, blow-through, air-swept, and side-entry.

The drop-through rotary feeder (Figure 3.12) is the original design of rotary feeder. Its body has a round, square, or rectangular cross section as well as inlet and outlet. It is used as an airlock below filter- and cyclone-receivers in vacuum systems, and as both airlocks and feeders for feeding materials into pressure and vacuum systems. As an airlock below filter- and cyclone-receivers, it should have a round inlet to maintain the circular figuration of the bottom cone to which it attaches. As feeders and airlocks for feeding materials into pressure or vacuum systems, either body shape can be used, but the rotor, in any event, should be normal to the centerline of the conveying pipeline.

The blow-through feeder (Figure 3.13) is used only for feeding finely ground, free-flowing, nonabrasive materials into low-pressure systems. The material enters the rotor through the top inlet. As the rotor pocket rotates to the bottom of the feeder, the conveying airstream literally punches the material out of the rotor into the conveying pipeline. The body of the feeder can be considered a part of the pipeline, making the feeder as a whole an integral part of the conveying pipeline. Designs are available for use with pipelines up to and including 6 in. in diameter. In choosing a feeder of this type, it is better to have it run fast than to have it run slow. A faster-running feeder will give a more uniform feed to the conveying stream. Also, the volumetric capacity of the feeder should be slightly lower than that for the

FIGURE 3.13.   Blow-through rotary feeder. (Courtesy Prater Industrial Products, Inc., Chicago, Illinois.)

conveying system so that each filled pocket is completely punched out into the conveying stream.

The air-swept feeder is a compromise between the drop-through and the blow-through types. It too is used for feeding materials into a low-pressure system. It does have the advantage over the blow-through type in that it can handle granular materials in addition to finely ground, free-flowing, and nonabrasive materials. Such a feeder also utilizes the conveying air to mix with the material in the feeder pockets, but its discharge is assisted by the influence of gravity, not by air velocity alone. Its volumetric efficiency is the highest in the rotary feeder field. Pipeline sizes used with this type of feeder vary from 2 to 6 in.

In the "Air-Swept" feeder (Figure 3.14), air under pressures up to 20 psig is directed into the feeder by air inlets in the hollow end bells at each end of the rotor shaft. From the hollow end bells, the air is directed through an opening (at approximately 7 o'clock) into a slot the full width of the feeder. From this slot, the air is directed into the rotor pockets, mixing with the material in the pockets, and discharging into the conveying pipeline through the feeder outlet.

A good feature of this feeder is that with air pressure in the hollow end

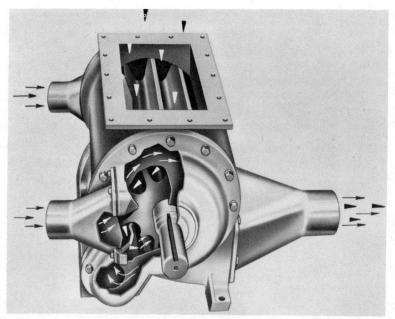

FIGURE 3.14. "Air-Swept" rotary feeder. (Courtesy CEA—Carter-Day Company, Minneapolis, Minnesota.)

bells and the shaft bearings located on the inside surface of the end bells, the bearings are free from contamination from the material being conveyed. An inside plate on the hollow end bell allows a slightly higher pressure in the end bell housings than the pressure inside the feeder, thereby creating an "air seal" around the shaft opening and periphery of the rotor.

The Dual Air-Inlet DA feeder (Figure 3.15) uses an adaptation of the solid headplate, plus the addition of integrally cast ducts in the plates. Air is admitted through these ducts, mixed with the material in the feeder pocket, and discharged with it through the bottom outlet either into shallow or medium-depth conveying pans, which connect to the conveying pipeline. Feeder efficiencies as high as 98.4% are attained by the combination of full gravity and pressure discharge. This feeder, with its gravity discharge, does not require all of the conveying air to pass through it. If a condition exists where such a feeder is deemed advantageous, but the air volume for conveying is too great for the feeder to handle, an intake tee can be used whereby part of the conveying air is diverted to the feeder and the balance directed into and through the intake tee.

If partial air sweeping of a feeder is necessary, it is possible to modify a drop-through feeder to gain the desired result. Air may be admitted into one of its headplates to partially fluidize or to give a pressure assist to the material to aid its discharge. Again, an intake tee is used with an air line to the feeder. This air line should run from the piping between the blower and intake tee to the feeder. Between this pipe connection and the intake

FIGURE 3.15.   Dual air-inlet rotary feeder. (Courtesy Fuller Company, Bethlehem, Pennsylvania.)

tee itself, a butterfly valve should be used to regulate the amount of air that is either diverted to the feeder or allowed to continue on to the tee.

The side-entry rotary feeder (Figures 3.16 and 3.17) is designed to be nonjamming when handling free-flowing cubes, pellets, chips, flakes, and other materials that tend to become pinched, sheared, or jammed in the conventional top inlet type feeder. These two feeders illustrate two approaches to achieve the same result.

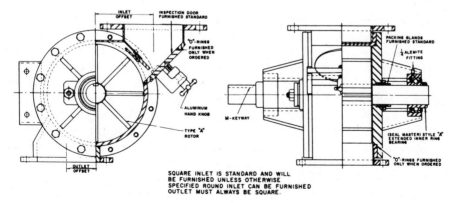

FIGURE 3.16.   Side entry rotary feeder. (Courtesy The Young Industries, Muncy, Pennsylvania.)

FIGURE 3.17.   Side inlet rotary feeder. (Courtesy Sprout-Waldron Division, Koppers Company, Inc., Muncy, Pennsylvania.)

The feeder shown in Figure 3.16 operates at a fixed speed and a maximum of 40% pocket full, with lesser throughput obtained through a slide adjustment near the throat of the feeder inlet. The feeder in Figure 3.17 operates at a variable speed to control throughput with the inlet especially designed to provide the necessary flow control.

In this matter of shearing and jamming, an orifice plate having a V-shape opening into the feeder is used at times to direct material into the rotor pockets. A simple baffle plate in the throat of a conventional top inlet can be used to good advantage (Figure 3.18). This is not a 100% guarantee against such occurrences, but it does an excellent job. It is also less of a restriction to material flow. It has been used very successfully in handling grains to minimize clipping of the kernels between the rotor blades and the body.

In the handling of fibrous materials and to maintain as large an intake opening as possible, the addition of shearing bars (Figure 3.19) to the feeder throat minimizes the jamming possibility. This feeder, designed to feed wood chips or any wood wastes into the airstream, indicates a thorough understanding of the particular needs when handling certain materials. Being a fabricated feeder, it is stress-relieved after fabrication, then machined. The housing, or body, is internally chrome plated to resist abrasion and corrosion; rotor blades have a lead angle or helix to produce positive shearing action; rotor tips are hard-surfaced stainless steel; a high-carbon steel knife at the intake throat shears "overs" and cleans the rotor blades to protect the housing from abrasion; and a bottom knife at the discharge opening cleans the rotor blades at that opening. Additional features include a hinged knife-access door with a safety interlock to prevent the feeder from

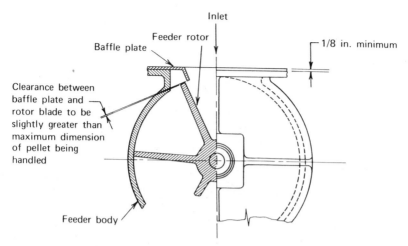

FIGURE 3.18.   Baffle plate for rotary feeder inlet.

operating when the door is open and replaceable body liners of chrome-plated mild steel easily replaced by only one row of studs.

To provide a rotary feeder that can be cleaned without complete removal, a cantilevered sanitary rotary feeder has been developed (Figure 3.20) for use in pharmaceutical, chemical, food processing, baking, and candy making industries. The central shaft has bearings at one end to provide a cantilevered shaft extension. The feeder housing is open on one end through

FIGURE 3.19.   Rotary feeder for wood chips. (Courtesy Rader Companies, Inc., Portland, Oregon.)

FIGURE 3.20. Cantilevered sanitary rotary feeder. (Courtesy The Young Industries, Inc., Muncy, Pennsylvania.)

which the rotor, rotor end seal, shaft seal, and spring-loaded pressure plate are assembled and disassembled.

For feeding abrasive materials into pressure-type systems, the Perma/flo* long-life solids feeder was developed. This feeder with superhard wearing parts and adjustability for wear has, according to the manufacturer, handled such materials as Portland cement, silica flour, pulverized limestone, feldspar, and like products. Acceptable maintenance is had when operating at lower pressure differentials, thus limiting abrasive wear due to the velocity of dust-laden air through fine clearances (air washing). Drives for this feeder include a pivoted gearmotor support and protection devices to stop and to restart the feeder in the event of an obstruction between the rotor and the body.

In vacuum systems, below the cyclone- and filter-receivers, one type of rotary feeder or discharge lock that should not be overlooked is the adjustable-for-wear. This feeder, good for vacuum service only, has its rotor carried in bearing arms that are adjustable by means of set screws. As wear takes place, the rotor can be lifted and adjusted to compensate for such

* Registered trademark, The Ducon Company, Fluid Transport Division, King of Prussia, Pennsylvanvia.

wear. The rotor is of the shrouded type. The body extends from the centerline of the rotor, or slightly below, up to the inlet flange. This means that sealing between the rotor and the body is limited to something less than 180°; that air sweeping due to minimum seal is more prevalent; and that the inlet neck of the feeder is drawn in from the flange down to its throat, which is a distinct disadvantage when handling fluid materials. Its bearings, however, are never in or near the product or material zone, thereby eliminating the need for any packing gland whatsoever.

All rotary feeders used in pressure service should be equipped with outboard, antifriction ball bearings. Where the rotor shaft extends through the headplates, packing glands of generous proportions should be used to seal against air or gas leakage into or out of the feeder during operation and to protect the bearings. If conditions warrant, and possibly require something better than a packing gland, a lantern ring should be added to the gland. The lantern ring will permit purging the glands continuously with air or gas to increase the life of the gland. In operation, the air or gas supply to the lantern rings need only be 1–2 psig higher than that in the interior of the feeder. In the author's experience, there was the case of an operator who complained that he could not maintain the packing in the glands of his rotary feeder, even with a lantern ring. On questioning, it was learned that he used plant air at 90 psig on the lantern rings. After reducing the air on the lantern rings to 12 psig when the system pressure was 10 psig, the problem vanished.

Vertical Feeders

Vertical feeders were used for about 60 years from their introduction around 1890 until their demise in the 1950s. They were used principally for feeding fibrous materials such as wood shavings, sawdust, and wood chips into low-pressure-type systems. The original feeder, invented by W. E. Allington in Saginaw, Michigan, consisted of a rotating vertical cylinder divided into five chambers by vertical partitions. The materials were introduced through an opening in a heavy cast iron top plate and carried around to register with openings in the top (air inlet) and bottom (air–material outlet) plates. The air from the blower was blown through continuously to clean out each chamber into the conveying pipeline. We might term it a vertical blow-through feeder.

In the mid-1920s, Guarantee Construction Co. of New York, through the efforts of two pneumatic conveying pioneers, P. R. Hornbrook and W. G. Hudson, sales and chief engineer, respectively, used the basic principle to design a similar large-capacity feeder for their Airveyor* ship and barge unloading systems handling grain. Their feeder used four 16-in. OD equally

---

* Airveyor was a trademark of Guarantee Construction Co. and since 1929, a registered trademark of Fuller Company, Bethlehem, Pennsylvania.

spaced pipes with top and bottom cast iron plates for the rotor instead of the vertical partitions of the Allington feeder. Two sizes were made with the pipes 24 and 36 in. high. The top air inlet was eliminated, since it was unnecessary for vacuum service. The grain was discharged by gravity through the bottom outlet, 180° opposite to the top inlet.

In the early 1930s, using what was thought to be the best of the two arrangements, vertical pipes and continuous air throughput, the feeder was used on wood chip systems using a higher pressure than that used by Allington. There were serious drawbacks to the feeder. Rosin, sawdust, and wood fines would build up on the stationary top, bottom, and rotating cast iron plates until the feeder would freeze. This meant taking it apart (no simple task); using a variety of cleaning methods, including planing, to remove the contaminents; and finally, reassembling it. Happily, the development of large horizontal feeders (Figure 3.19) made the vertical feeder obsolete.

Gate Locks

The three-compartment gate lock (Figure 3.21) is used as a discharge lock below a cyclone- or filter-receiver in a vacuum system and as a feeder to feed material into the pipeline system of a pressure system. In both applications it is more an airlock than a feeder. It is used for handling abrasive materials as well as materials that have a tendency to build up on the interior of rotary feeders.

The lock for pneumatic conveying consists of three gates, two of which are airtight. Only the middle and lower gates need be airtight. The upper gate is used as an interrupter to the flow of material into the lock. A continuous flow of material through the lock cannot be maintained, since each gate serves to lock a compartment, very similar to a batching arrangement. As each gate is opened, a batch of material discharges from the compartment above to the one directly below. To maintain either the vacuum in a vacuum system or the pressure in a pressure system, the airtight middle and lower gates open and close alternately so that when one is open the other is closed.

Since this lock cannot absorb a continuous flow of material per se, a continuous flow through a metering feeder and a small surge hopper or space above the top interrupter gate allows the unit to be used under such conditions (Figure 2.7). It cannot, under any conditions, be used under a head of material, such as a bin or hopper, where such head is greater in volume than the volume contained in any individual compartment of the lock.

Two sizes are available with maximum discharge capacities of 9 and 17 cu ft of material per minute, respectively. Maximum operating pressure is 15 psig. Operating temperatures are 250°F using urethane as the sealing

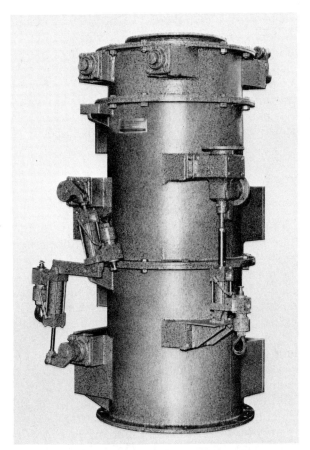

FIGURE 3.21.   Gate lock. (Courtesy Fuller Company, Bethlehem, Pennsylvania.)

material on the gates and relief valve sealing disks and 400°F using Viton*
rubber as the sealing material.

Developed primarily for fine, abrasive fly ash conveying, the PERMA/
lok III† abrasion-resistant airlock is primarily a gate lock in operation. It
too requires a metered feed when used under a head of material. It can also
be used both as a feeder for feeding material into a pressure-type system
and as an airlock under a filter-receiver of a vacuum system. Maximum
operating pressure is also 15 psig. Its capacity is variable by inserting cylindrical sections between the chambers and by varying the cycling frequency
up to a maximum of about 4 cpm.

---

* Registered trademark, E. I. duPont de Nemours & Co., Inc., Wilmington, Delaware.
† Registered trademark, The Ducon Company, Fluid Transport Division, King of Prussia,
Pennsylvania.

Air sealing between the chambers is provided by a stainless steel disk sliding into position over a gem-hard ceramic ring seat. The standard cast iron model can handle materials up to a temperature of 700°F. Higher temperatures are dependent on the materials of construction.

With both feeders, when a high volume of material is fed into a pressure-type system, care must be exercised in the size and design of the air–material intake to avoid a plugged system. Either a system large enough to absorb the volume of material or a metered feed into the pipeline system will be required.

Fluid Solids Pump

The two types of fluid solids pumps available are categorized in the low and medium type of pressure systems. They all are limited to the handling of materials that are fluidizable, with all the material being handled passing through a 50 mesh sieve and the bulk of the material passing through a 200 mesh sieve. The material must also be conducive to flow, not to compact. If you squeeze a handful of material in your hand and it oozes out between your fingers, it generally can be handled with these pumps. If the material compacts, however, and if you have in your hand a snowball, its chances for handling are extremely limited.

In the medium-pressure field, the Fuller-Kinyon Pump operates under air requirements of 15–45 psig, depending on the capacity or conveying rate, and the conveying distance of the system. Since this Fuller-Kinyon Pump played and continues to play an important part in the pneumatic conveying field, let us review its details of construction and operation (Figure 3.22). All parts of a Fuller-Kinyon Pump are mounted on a cast iron base (1). The materials to be conveyed enter the hopper (2) by gravity from the usual sources of supply, and are advanced through the barrel (3) by the impeller screw (4), which is directly driven through a flexible coupling connection to the driving motor. As the material advances through the barrel it is compacted by the decreasing pitch of the impeller screw flights, and it is further increased in density by the space or seal between the terminal flight of the impeller screw and the face of the check valve disk (5). The exact density required is further controlled by an adjustment of the seal length by means of the jack screws (6). The material then enters the check valve body or mixing chambers (7), where it is made fluent by compressed air introduced through a series of air jets (8), and from there is enters the transport pipeline. The pump impeller screw is supported in a hollow shaft (9), which in turn is supported by the ball bearings (10 and 11) in a single bearing housing (12). The supporting hollow shaft is rotated in the bearings by the impeller screw shaft by means of an eccentric lock collar (13). The impeller screw (4) can readily be removed through the check valve body or mixing chamber by removing the cover plate (14) and loosening the lock collar (13).

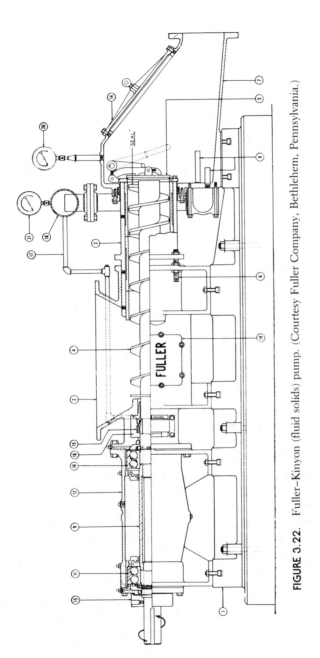

FIGURE 3.22. Fuller–Kinyon (fluid solids) pump. (Courtesy Fuller Company, Bethlehem, Pennsylvania.)

There are no packing glands in the Fuller-Kinyon Pump. The material in the hopper is sealed from the bearings when the pump is in operation and under pressure by means of an air-cooled mechanical seal ring (15) in the chamber (16). The seal ring is kept clear of material from the hopper by means of the purge air supplied through air piping (17) from the header (18). Ample clearance is provided for the impeller screw in the hopper section. This hopper section has a catch basin on the bottom for tramp iron, and a clean-out door (19) on each side. The barrel is protected by renewable, wear-resistant liners, and the screw flights are also protected with a special alloy to give maximum service. The air supply enters the air manifold, which supplies the air jets (8). The transport line pressure is indicated on the gauge (20); another gauge (21) shows the air-jet pressure, which with medium pressure operation is but a few pounds higher than the pressure shown on the first gauge (20). The conveying pipeline is cleaned or blown out by stopping the impeller screw and blowing air through the air jets for a few minutes. The check valve (5) in this case prevents the air from blowing back through the pump.

Air pressures and volume requirements of Fuller-Kinyon systems are proportional to the quantity of material handled and the conveying distance. Since the cost of compressing air increases rapidly as the pressure increases, it is desirable to limit power input to the actual conveying workload at any time. Rotary air compressors allow this to be done by delivering a uniform flow of air, without pulsations, directly into the Fuller-Kinyon Pump, without unloaders or air receivers. Connected in this manner, the power input conforms automatically to the workload (back pressure) of the conveying system.

It must be remembered that the fluid solids pump is, for all practical purposes, a volumetric feeder. In other words, for a given size pump, its rated capacity is dependent on the bulk density of the material: light materials, low capacity; heavy material, high capacity.

As outlined in Chapter 2, the design of complete systems using fluid solids pumps must be left with the manufacturer. Practical knowledge of their capabilities are the manufacturers' main forte.

The pumps are manufactured in various sizes. For handling materials in the 30 lb/cu ft range, their capacity will range from 2 to 80 tons/hr. The pipeline sizes will range from 3 to 10 in., with the maximum conveying distance going around the 2000-ft mark. The power for the pump and air supply will range from 17.5 to 250 hp, and the maximum air pressure will be 25 psig. For materials in the 70 lb/cu ft range, the pump will have a capacity range of from 6 to 150 tons/hr, with the pipeline ranging from 3 to 12 in., the maximum conveying distance 2000 ft, and the power required for both the pump and the air supply ranging from 17.5 to 400 hp. In this category, the maximum air pressure will be 30 psig. For materials in the 100 lb/cu ft range, the pump will have a capacity ranging from 6 to 300 tons/ hr, with the pipeline sizing from 3 to 12 in. Again, the maximum conveying distance will be roughly 2000 ft. In all these cases, however, by conveying at reduced rates for the same-sized pump, conveying distances of 4000 ft are possible. For the 100 lb/cu ft material, the power for both the pump and the air supply ranges from 25 to 1250 hp, and the maximum air pressure will be 40 psig.

The latest design of this pump features support at both ends of the screw impellers and a side discharge. Four sizes are available from 150 mm (5.9 in.) to 300 mm (11.8 in.) to provide conveying rates of up to 450 metric tons (495 U.S. short tons) per hour. The greatest advantage of this latest design over the original is the ability to vary conveying rates through adjustments to the speed of the screw impeller.

Another fluid solids pump, the KVS-Möller, operates on somewhat the same principle as the Fuller-Kinyon. The patented variable-pitch impeller screw, however, can handle powdered, granular, or pulverulent materials with variation in conveying rates without changing the speed of the impeller screw. The discharge from the variable-pitch impeller screw terminates in a mixing chamber containing an air jet that is normal to the centerline of the screw. The conveying pipeline can radiate in all directions, again, normal to the impeller screw.

The Peters Pump is a fluid solids pump patterned very much after the Fuller-Kinyon, but with constructional variations. This pump can operate against pressures of up to 70 psig with conveying rates of up to 500 tons/hr over a conveying distance of 3300 ft.

The Fuller-Kompact Pump is basically an adaptation of the Fuller-Kinyon Pump simplified to operate with no greater air pressure than 12 psig and a maximum conveying distance of 125 ft. The pump has a capacity of 54 tons/hr maximum when conveying Portland cement, uses conveying pipeline sizes of 4, 5, and 6 in., and uses from 40 to 100 hp to operate both the pump and the air supply. It was designed specifically for conveying requirements of relatively short distance, intermittent or light duty, and high capacity in terms of pump size. Two types are available: one to meet the needs of plants that receive materials by truck or railcars; a second for plants that reclaim from ground storage and elevate to a batch bin or other receptacle.

Blow Tanks

The blow tank is basically a tank filled with material and emptied into a pipeline by the energy of expanding compressed air admitted to the tank either singly or in pairs, using a combination of fluidization, pressure balance, and/or controlled acceleration, depending on the material to be conveyed. There are many who believe that only dry materials can be handled. This is not so. The pneumatic ejector for handling screenings and grit in water and wastewater treatment plants has been in use for nearly 50 years. In more recent years, the ejector has also been used for handling semidried solids and wet, free-flowing slurries.

In handling dry materials, until the application of the blow tank principle to the pressure differential method employed in self-unloading bulk trucks and railroad cars, its use was limited to the handling of pulverulent materials. With the application of the pressure differential method, materials up to 1 in. in size have been conveyed.

The blow tank can be categorized as a universal pressure-type system since it is designed today to operate under low-, medium-, and high-pressure air requirements. The low-pressure systems are classified by the use of rotary blowers as their supplier of the air requirements up to 15 psig. Above 15 psig we can consolidate the medium- and high-pressure systems into a single category, since they both must use heavier-duty air compressors, air receivers for high pressure to satisfy total air requirements, and the tank must be built to conform to the ASME pressure vessel code and to meet state and local regulations.

In the handling of pulverulent and fluidizable materials, as the material enters and fills the tank to a predetermined level (not water full), the material will deaerate and pack to a certain extent. Before conveying can be started, the material must approach the fluid state. This fluidization is accomplished by admitting air into the material at or near the discharge outlet of the tank, which is the starting point of the conveying pipeline. As this air permeates the material making it fluid, it also builds up pressure within the tank to the point where this pressure forces the material out of the tank and into the conveying pipeline. The operating (tank) pressure will be in the neighborhood of 2–4 psig higher than the pressure in the conveying pipeline. Rate of flow out of the tank and into the conveying pipeline is governed by control valves and air volume and pressure variations.

In the handling of coarse materials, there are usually sufficient voids between particles to allow complete pressurization of the tank and material without the necessity for special aeration equipment. The tank is pressurized first, then through regulating valves the pipeline is pressurized to about 2 lb less than the tank. Again, pressure differential causes the material to flow from the tank into the pipeline and on to its destination.

Later developments (Figure 3.23) have improved on the original pneumatic ejector whereby the blow tank is filled with material, then high-pressure air is directed into the tank near or above the top level of the material in a manner that pressurizes the tank to force the material out through the bottom outlet and into the conveying pipeline. The material then traverses the pipeline as a coherent slug to its destination, controlled by an air control device that continually modulates the air pressure and volume to ensure a constant low velocity.

While all blow tanks operate under the aforementioned methods, there are many types and capacities available. They all use a low ratio of air to material. In the low-pressure range they vary in size from 3 to 1000 cu ft. In the medium- and high-pressure range, they vary in size from 6 to 700

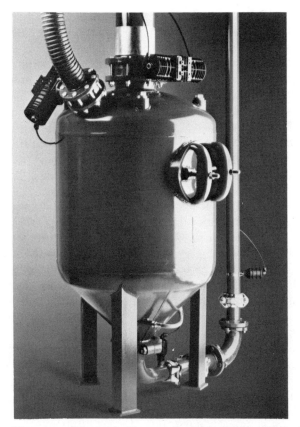

FIGURE 3.23.   Ejector-type blow tank. (Courtesy Dynamic Air, Inc., St. Paul, Minnesota.)

cu ft. Conveying rates are similarly variable, pipeline size and length being the prime factor.

For the design of the system, air requirements, pipeline size, and general operation, consult the manufacturers supplying this type of equipment. The experienced ones have a wealth of proprietary and operating information on which to draw to arrive at their recommendations. In discussing the overall operation with them, be certain that operating sequences are understood by all concerned. Timing in this system is most important: filling, fluidization or pressurization, emptying and conveying, and recycling all consume time. The overall capacity, not the conveying rate, is governed by the time each cycle or batch consumes from the initial filling of the tank to completion of emptying and conveying, and return to filling, together with the volumetric capacity of the tank.

When conveying to terminal points where dust suppression equipment is employed, be extremely careful in the sizing of the dust suppression equipment. Be liberal, not conservative. Although the quantity of the normal conveying air might be small, when the tank becomes empty a so-called bubble of compressed air is released through the pipeline system. This bubble of air, although momentary, can cause difficulties with marginally designed dust suppression equipment. In small blow tanks and low-pressure air, this bubble may not be of much consequence, but in large blow tanks and high-pressure air, the air volume at atmospheric conditions can be quite high. When the air in a 400-cu-ft-capacity blow tank at 30 psig is expanded to 14.7 psia, its volume increases to approximately 1200 cu ft, and at 60 psig it increases to approximately 2050 cu ft. At times, it must pass through the dust suppression equipment within a matter of a few seconds. Baghouse or dust collectors should be sized accordingly.

Blow tanks are economical in operation, using minimum air requirements; they are efficient, since during the conveying operation no air is lost through shaft seals; they are dust-tight, minimizing a dusty atmosphere and working conditions; and they are excellent in the handling of heat-sensitive materials, since no moving parts are ever in contact with the material being conveyed. One word of caution, however: Do not use overheated air. Air in blow tank systems is in contact with the material a longer period of time than in low-pressure-type systems.

Piping

Piping used as pipelines for pneumatic conveyors handling most materials consists mainly of standard steel, IPS, Schedule No. 40, in sizes 2 to 7 in. in diameter, and Schedule No. 30 in 8 to 12 in. in diameter. Piping of 14-in. diameter and above is welded steel pipe in wall thicknesses varying between $\frac{3}{16}$ and $\frac{5}{16}$ in., depending on the material being conveyed. For conveying plastics and other materials that require contamination-free conveying, aluminum and stainless steel pipe is used, also in IPS sizes, but in Schedule No. 5 and No. 10 wall thickness.

When a change in direction is required in the conveying pipeline, long radius bends should be used. In conveying most chemicals and plastics, these bends are made up of the same size pipe as that used for straight tangents. Wherever possible, the rule of thumb of 12 in. of radius for each inch in diameter of the pipeline should be used, up to and including 8 in. Above 8 in., the recommended radius should remain at 8 ft, except when using 14 in. and above, where larger radii should be used. Table 3.14 indicates the recommended radius as well as an allowable minimum radius.

Pipe bends are fabricated either by cold bending in a pipe bending machine or by heating and then bending in a forming device. In cold bending, the pipe is guided by rollers formed to fit the outside diameter of the pipe. The rollers are positioned on the inside and the outside of the pipe bend as it is run through the machine. This causes the pipe to be flattened to a small degree as it is bent. In other words, the cross section of the pipe in

TABLE 3.14.   Pipe Bend Radii

Pipe Size	Minimum Radius	Recommended Radius
3 in.	2 ft, 0 in.	3 ft, 0 in.
3½ in.	2 ft, 3 in.	3 ft, 6 in.
4 in.	3 ft, 0 in.	4 ft, 0 in.
5 in.	3 ft, 6 in.	5 ft, 0 in.
6 in.	4 ft, 0 in.	6 ft, 0 in.
7 in.	4 ft, 6 in.	7 ft, 0 in.
8 in.	5 ft, 0 in.	8 ft, 0 in.
10 in.	6 ft, 0 in.	8 ft, 0 in.
12 in.	6 ft, 0 in.	8 ft, 0 in.

the curved portion of the bend becomes slightly oval, compared to the circular cross section of the straight pipe. In hot bending, the pipe is filled with sand, then heated and bent. This method retains much more the round cross section. In any bend, crimping of the pipe wall is to be avoided at all cost. Crimping is a detriment to the operation of the conveying system.

Plain (nonreinforced) pipe bends are used for nonabrasive and slightly abrasive materials. To prolong the life of the bend in slightly abrasive materials use attached reinforcements. These can be either a flat steel plate formed to follow the contour of the pipe bend closely at its outer surface and continuously welded to the pipe, or segmented castings of either cast or white iron that are both cemented to the back of the pipe bend and clamped on through U-bolts. In using the segmental casting method, the pipe bend should be bent hot. For highly abrasive materials, wearing boxes, made up of steel plates, are welded to the back of the pipe. These boxes can remain hollow or can be filled with an abrasion-resistant concrete. If a hollow box is used, when sufficient wear to the pipe itself results in a hole, the hollow space will fill up with conveyed material. When it is filled up and wear results in additional holes in the pipe, the material being conveyed will impinge on itself. If the material is inert, this is all right, but if the material is friable, use caution. When the box is filled with concrete it will be most resistant to abrasion. If concrete-filled wearing boxes are used, do not fail to take into account the weight of the concrete in the bends when designing and spacing the necessary supports.

A comparatively recent development is the use of centrifugally cast ductile iron pipe. This pipe is used principally for conveying highly abrasive materials such as fly ash. Straight lengths are available up to 18 ft long; wall thicknesses are available from ⅜ in. for 2-in. ID to ¾ in. for 12-in. ID. Bends are available with integral and replaceable wearbacks. Integrally cast wearback bends include 22.5°, 30°, 45°, 60°, and 90°. Replaceable wearback bends include 45°, 60°, and 90°. Radius of these bends for all sizes, 2–12 in., is 3 in. for each inch in diameter.

In the material specifications, the item of abrasiveness of the material to be conveyed was classified according to numbers of the Moh Scale. We can here classify the type of pipe bend according to the scale as follows:

Moh Scale	Type of Bend
1, 2, 3	Plain pipe
4, 5	Reinforced pipe
6, 7	Pipe plus wearing boxes

Certain materials require bends other than standard pipe and reinforcements. When handling material such as alum, which has a tendency to adhere to pipe bends, the bends should be made of rubber hose, properly reinforced to withstand the required vacuum or pressure without collapsing, and static electricity dispersion wires. The hose should not be allowed to drape free but should be supported by a structural member, formed to the proper radius for the size of pipe used.

In handling grains, precautions should be taken to minimize breakage to the kernels as well as to avoid skinning the jackets of the kernels. A point of extreme caution is the pipe bends in the conveying pipeline. Grains are abrasive and cause wear to standard pipe bends. This wear is not uniform throughout the cross section of the pipe. The pipe wall will wear in rivulets. As these rivulets are formed, they are arranged in a most irregular manner, and they have rough edges. As the grain is impinged against the inside of the pipe with the rivulets, the kernels are damaged and the hulls skinned. To avoid this condition, bends are made up of steel plate having a rectangular cross section; further, an abrasion-resisting renewable wear plate of no greater thickness than $\frac{3}{16}$ in. should be used for the back of the bend. The rectangular cross section should maintain the inside diameter of the conveying pipeline for its width. Its depth is to be such that the net area of the bend should be equal to the area of the conveying pipeline. Transitions from round to rectangular, and from rectangular to round, are to be provided at the points of curvature and tangency of the bend. At the point of tangency, the interior of the transition from the back of the bend to the round pipe should be a straight line. The length of the transition from round to rectangle and rectangle to round should be about six times the diameter of the conveying pipeline to provide an easy flow with minimum turbulence.

Pipeline joints must be as tight as possible, and they must be airtight. There should be no gaps between the ends of the abutting pipes. Joints include flanged, welded, and coupled. Flanged joints should have the pipe ends flush with the face of the pipe, and a gasket of minimum thickness. Welded joints can be either strap or butt welded, but care must be exercised in welding so that welds do not protrude into the inside of the pipe. Coupled joints are used in vacuum and low-pressure systems. The couplings are of the compression type known as "Dresser" couplings. This and similar couplings will make a tight joint; they will take up some expansion and contraction, and they will permit some flexibility at each joint. For pneumatic conveying, the pipe stop in the center of the coupling's sleeve must be removed so that the pipe ends can butt. In fabricating the piping, the ends of the pipe must be true and square with long axis of the pipe. Also, burrs on the inside should be removed.

Any discussion of piping would be incomplete without mention of one of the innovations used in fluidized dense-phase low-velocity conveying. All systems of this type feature medium-pressure blow tanks as the prime mover with the material conveyed as a dense slug at low velocity. Inline boosters are used where the conveying pipeline pressure rises, indicating a possible line plug forming, to add a blast of compressed air to the piping system to give the material a boost to keep it moving. The booster or antiplug device is built around a specially modified diaphragm relief valve that senses increased line pressure to the point where the valve opens, giving the blast of air to the pipeline until the line pressure returns to normal.

These boosters are also used in elbows of various configurations with the elbows replacing long-radius bends. The elbows also use a dead space so that the material being conveyed impinges on itself, rather than on the internal walls of the elbow. This same principle was used some 40 years ago, but under different conditions. A high-velocity vacuum system handling record scrap (a mixture then of shellac, carbon black, and various fillers) used a cast iron long-radius tee at the top of a vertical riser to change direction to horizontal. The long-radius sweep was the off-leg of the tee with the straight through leg capped. In operation the material filled the straight-through leg, also allowing the material to impinge on itself.

Another innovation is the booster fitting placed at intervals along the pipeline system to overcome frictional resistance losses and to prevent plugging. The amount of air added to the air–material mixture in the pipeline is often minimal. Other times it can be substantial. The design of the pipeline, including any type of booster, should be made carefully, especially the number and placement of the boosters.

Rather than boosters, a bypass pipe with equally spaced holes is included within and as part of the conveying pipeline to minimize plugs by allowing the conveying air to detour around a plug, to reduce its leading edge so that the plug disappears. Materials for this arrangement of piping should have low permeability and high air retention.

As part of the pipeline, flexible hoses, whether used as pipe switches or in vacuum car unloading systems, should receive equal consideration. The type of required hose, metal or rubber, is wholly dependent on the material handled. Normally, nonabrasive and mildly abrasive materials are handled in flexible metal hoses, and abrasive and sanitary required materials are handled in rubber hoses.

Flexible metal hoses are used mostly in car unloading operations. Under such service, they are subjected to much manhandling and operation under all types of conditions. For longest service and freedom from corrosion and rust, a double-wall stainless steel hose with a smooth inner liner is recommended.

Rubber hoses for such service should also have a smooth inner liner, or what we call a "smooth bore." The hose should be heavily constructed and to be reinforced with wound static dissipating wires that are fastened (welded or brazed) to the end metal couplings. Without static protection, the hose would look like the worst possible case of chicken pox. Also, garden-type-hose construction should never be used.

For handling food products where sanitary construction and no impartment of taste to the material being conveyed are necessary to meet FDA requirements, type D_2, food grade natural rubber should be used as the inner liner material. These hoses too should be wire reinforced and static conductive.

When designing a pipeline system, avoid any and all constrictions in the internal movement of material. Pipeline and other internal areas, including those handling air only, should be their smallest at the intake point of the system. Any change downstream should enlarge the area, not reduce it. A good illustration is the design of an intake manifold (Figure 3.24) where two 5-in. car unloading hoses are attached to an 8-in. conveying pipeline.

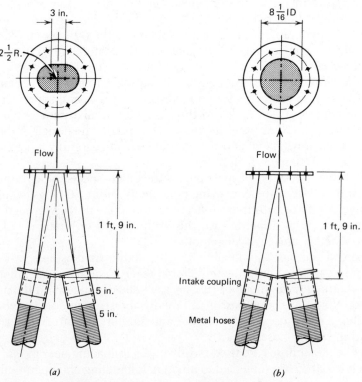

FIGURE 3.24. $8 \times 5 \times 5$ in.^{3} intake manifold.

Figure 3.24a was designed, manufactured, and installed in a vacuum-type unloading system. The two 5-in. hoses had a total area of 39.27 sq in. By simple expedient, the designer just halved a 5-in. pipe and welded two side plates to form the enclosure with an oblong opening in the top flange, 5 in. wide by 8 in. long. Its area totaled only 34.64 sq in. This meant that the material coming out of a total area of 39.27 sq in. had to squeeze through an area of only 34.64 sq in. In a system designed to unload at a rate of 25 tons/hr, all that could pass this restriction was 20–21 tons/hr. Replacing that manifold with one of proper design (Figure 3.24b), increasing the opening in the top plate to match the ID of the 8-in. pipeline (area of 51.05 sq in.), the conveying rate immediately went up to 26 tons/hr.

Pipe Switches

Pipe switches, or diverting valves as they are frequently called, are used in the conveying pipeline to divert the conveying stream from a single pipeline to two or more separate pipelines. Conversely, they are also used for conveying from two or more lines to a single one. Divergers are used mostly in pressure systems, whereas convergers are used mostly in vacuum systems. One paramount fact describes the difference between a diverting and converging valve in their respective systems. In a pressure system, the valve and its housing should be as airtight as possible so that leakage is kept to an absolute minimum. In a vacuum system, the valve need not be as airtight, but the leakage still should be kept to a minimum.

In pressure systems, any leakage not contained within the pipeline system will allow dust to escape to the atmosphere. In closed rooms or buildings, and even outdoors, the escaping dust can be not only a nuisance but also a perplexing situation. Working conditions (dusty atmosphere) as well as a dirty plant wherever the dust settles will be the result. Diverter valves in pressure systems must have airtight housings or, if housings are not used, tight connections.

In vacuum systems, leakage not contained within the pipeline system is not too serious a problem to the adjacent atmosphere, since leakage is in, not out. Leakage, however, should be kept to a minimum since whatever leakage does occur adds only to the amount of air to be handled by the exhauster above that required for conveying.

In the early days of pneumatic conveying, pipe switches were simple, unsophisticated affairs. The first switch used a flexible element permanently connected to a single pipeline connecting to the two or more divergent pipelines through make-break couplings (Figure 3.25). In today's terminology, it might be described as a crude mechanism. It was and still is a complete success, being tight under both vacuum and pressure service. When installed in a pipeline it does not require a long hose flexible section to accommodate the required angularity, and it does not require longitudinal movement to make and break the joint. Through the toggle action of the

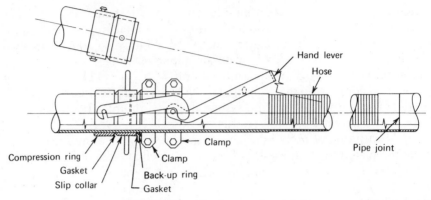

FIGURE 3.25. Toggle-type pipe switch.

hand lever, power is applied to the compression ring to compress the gaskets on each side of the slip collar, making the make-break joint airtight. The flexible element is made up of flexible metal hose having a hardened steel overlapping liner for material classed as mildly abrasive, and for abrasive materials a rubber hose with a smooth bore is used. As a manual switch, this unit has no peers. It is, however, not adaptable to power and automatic operation. Today, in this type of switch, couplings such as the Kamlok* are substituted for the make-break coupling. In using such couplings, the flexible element must be much longer so that longitudinal movement can be accommodated to allow the male part of the coupling to be removed or inserted, with little or no resistance, into or out of the female part.

Taking the same principle of the flexible hose, there next came the rack and pinion pipe switch for vacuum service. This unit operated through an arc of about 20° per pipeline connection. It also eliminated the make-break connection. Through the rack and pinion, the switch was moved to the desired locations by the turning of a handwheel. Normally, the maximum number of pipes connected through the switch was three on one side and one on the other. This type of switch was adaptable to either air or electric motor operation. Using the rack and pinion principle and longer flexible hoses, pipe switches up to eight-way have been made. With the addition of power closures to the movable portion of the switch, it became a very useful switch for pressure service. These switches were expensive in their manufacture, which led to the enclosed type of switches and diverter valves available today. They should not be entirely forgotten, however, since the curved flexible portion of the switch is much more gentle to the handling of grains and friable materials than the more abrupt change in direction employed in today's switches.

As the use of the fluid solids pump was expanded some 60 years ago, the need for a diverter valve that would be completely airtight and void of leakages up to 40 psig led Smith and Kintner to develop the S-K Valve (Figure

* Registered trademark, Dover Corporation/OPW Division, Cincinnati, Ohio.

3.26). The basic design of this valve has withstood the test of time. Its basic shape and operation have remained unchanged, improvements being required only in materials of construction to better withstand abrasion. It is used primarily in systems of the medium-pressure range handling fluidizable materials. It has been used in low-pressure systems where rather high velocities are needed, even though the resistance through the valve is much greater due to the higher velocity of the conveying stream, and the turbulence in going both straight through and to the off-leg. Two-way valves are constructed for either manual or electric motor operation, and in sizes of 3–12 in. In three-way operation, they are constructed for manual operation only, and in sizes 4–10 in.

Since World War II, the great increase in the number of manufacturers of pneumatic conveying systems and component parts has brought forth probably as many types and arrangements of pipe switches or diverter valves as there are manufacturers. They all involve either the flap type, plug type, flat slide type using flexible elements (an offshoot of the rack and pinion curved slide), or the truncated conical-shaped rotor that rotates within its cylindrical housing. The choice of the diverter valve to use depends on the material being handled. On the enclosed valves, the flap type is normally used for powdered or finely divided materials, whereas the plug type is used for nondusty, pelletized materials. The flat slide diverter and conical-shaped rotor type can be used for all types of materials. Most of these diverter valves are manufactured in two-way only and can be operated manually or through air or electric motors. Materials of construction usually used are cast iron, aluminum, and stainless steel, with the choice being governed by the material being conveyed. Operating pressures approach the 20 psig range and apply for pipeline sizes 2–12 in.

When using air cylinders or motors to operate the diverter valves, care should be exercised in their location in the installation. They should not be used where below-freezing temperatures occur because air lines, solenoid-operated control valves, and cylinders and motors will freeze because of moisture in the air lines. One client in central New Jersey had a pneumatic installation handling PVC with numerous air-operated diverter valves and pipe switches. Needless to say, to keep this operating during the winter, all the valves as well as the air lines themselves were wrapped with electric heating cable to keep things from freezing up. If valves are used outdoors or in unheated buildings where the temperature will go below the freezing mark, consider electric motor operation. Electric wires and motor windings do not freeze. If you feel you must have air operation, be sure dry air is available at all times.

Air-Activated Gravity Conveyors

The air-activated gravity conveyor, considered a separate conveying unit in comparison with a pneumatic conveyor, here will be considered a component and an important accessory to the complete pneumatic conveying

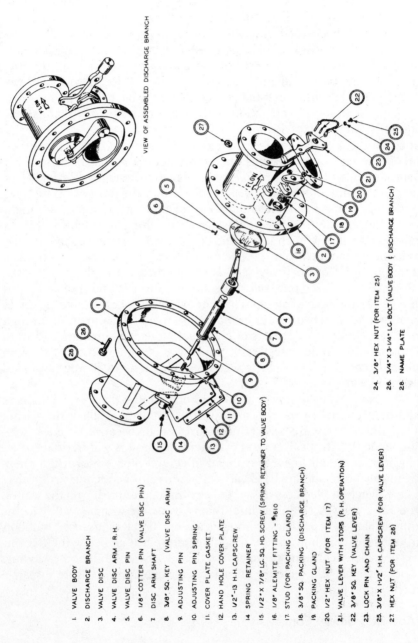

VIEW OF ASSEMBLED DISCHARGE BRANCH

1. VALVE BODY
2. DISCHARGE BRANCH
3. VALVE DISC
4. VALVE DISC ARM - R.H.
5. VALVE DISC PIN
6. 1/8" COTTER PIN (VALVE DISC PIN)
7. DISC ARM SHAFT
8. 3/8" SQ. KEY (VALVE DISC ARM)
9. ADJUSTING PIN
10. ADJUSTING PIN SPRING
11. COVER PLATE GASKET
12. HAND HOLE COVER PLATE
13. 1/2"-13 H.H. CAPSCREW
14. SPRING RETAINER
15. 1/2" X 7/8" LG. SQ. HD. SCREW (SPRING RETAINER TO VALVE BODY)
16. 1/8" ALEMITE FITTING - #1610
17. STUD (FOR PACKING GLAND)
18. 3/8" SQ. PACKING (DISCHARGE BRANCH)
19. PACKING GLAND
20. 1/2" HEX NUT (FOR ITEM 17)
21. VALVE LEVER WITH STOPS (R.H. OPERATION)
22. 3/8" SQ. KEY (VALVE LEVER)
23. LOCK PIN AND CHAIN
24. 3/8" HEX NUT (FOR ITEM 25)
25. 3/8" X 1-1/2" H.H. CAPSCREW (FOR VALVE LEVER)
26. 3/4" X 3-1/4" LG. BOLT (VALVE BODY & DISCHARGE BRANCH)
27. HEX NUT (FOR ITEM 26)
28. NAME PLATE

FIGURE 3.26. S-K valve. (Courtesy Fuller Company, Bethlehem, Pennsylvania.)

system. It is also used in conjunction with mechanical conveying systems, not as elevating conveyors, but as horizontal components of the complete system.

As described in Section 2.9, this conveyor operates on the ability to fluidize material through controlled aeration so that the material reacts similarly to water. In their fluid condition the material seeks its own level. With the conveyor on a slight slope (normally 2–6° for most materials, but steeper for some others), the material, seeking its own level, flows down the declining slope.

The porous membrane is the heart of this conveyor. Most prominent are the woven cotton and polyester fabric and the porous plates (ceramic and aluminum oxide). No attempt will be made here to discuss their relative merits. They are both used very successfully, with possibly the woven fabric enjoying greater acceptance. Other materials include plastics, fiberglass, microporous stainless steel laminates, and sintered stainless steel. In any case, the rules that need to be followed in design and use are practically the same. This is evidenced by the capacities in cubic feet per hour for woven fabric and porous plates (Table 3.15) for closed-top conveyors with the size being the internal width of the net conveying area. When necessary, cotton fabric can be treated for mildew resistance and for fire retardance.

In handling materials having elevated temperatures, the woven cotton fabric will withstand a temperature of 275°F for a prolonged period of time. With the addition of a combination of asbestos to resist heat and wire cloth to resist abrasion, the temperature of the material being conveyed can approach 550°F. The plastic membrane will handle materials up to 200°F; the polyester membrane, up to 350°F; the ceramic porous membrane in standard construction, up to 300°F and in special construction, up to 800°F. The stainless steel membranes, sintered or laminated, and fiberglass will handle materials up to 800°F. The stainless steel membranes are especially

TABLE 3.15.   Capacities—Air-Activated Gravity Conveyor

Conveyor Size (in.)	Woven Fabric (cu ft/hr)	Porous Plates (cu ft/hr)
4	460	—
6	1,200	1,200
8	2,500	1,800
10	3,500	3,000
12	6,000	4,500
14	8,000	6,500
16	10,000	10,000
18	—	14,500
19	16,000	—
20	—	20,000
24	22,250	26,000

applicable where all stainless steel construction is mandatory for the material being conveyed.

The length of a single conveyor can approach a mile. All that is needed is headroom to maintain the slope throughout its length and sufficient air to maintain fluidization.

Dependent on the placement of the conveyor in the overall scheme, flow regulation down the conveyor may be absolutely necessary—but no flow regulation at all may be the answer too. When using the conveyor to take material directly from the outlet of a bin, or some such large volume, if no regulation to flow is made, the material could very well gush down the conveyor, making control, either at the end of the conveyor or beyond, a very difficult operation. If the material is received by the conveyor from a metered or regulated flow, say from a bucket elevator or other similar piece of machinery, no flow regulation should be used. If the flow from the preceding machinery were to surge and deliver much more material than a regulated flow could handle on the conveyor, backup of material could occur in that piece of machinery. This could cause a serious breakdown all along the line.

Accessories for this air-activated gravity conveyor include flow-regulating gates used as outlined earlier, cutoff gates, Y-pieces or stream splitters to split the flow from one conveyor to two or three and vice versa, side discharge boxes that will allow discharge of the material anywhere between the inlet and the outlet, and material traps. Flow-regulating, cutoff, and side discharge gates can be furnished for either manual or automatic operation. Inspection ports or doors should be placed or used in the top conveyor section at the inlet to the conveyor, after each accessory, and at the discharge. With the conveyor being a completely enclosed unit, the observation ports enable an observer to see what is happening inside the conveyor without resorting to either partial or full dismantling.

The material trap is a collection agent for tramp iron and heavier foreign materials than the fluidized material. It is a small section inserted into the conveyor that simulates a small bin that has completely fluidized material in it. As the heavier particles reach this section, they fall by gravity into the bin. Of course, this unit must be emptied periodically or it will fill with tramp material and, in this condition, collect no more.

Air requirements for these conveyors vary considerably. Different materials have a large variety of densities and finenesses. Their moisture content will have a large bearing on the ability of the material to fluidize. Moreover, some materials of extreme fineness, instead of remaining fluid under the influence of air, will agglomerate. When this happens, the air-activated gravity conveyor just will not operate. Unless the identical material with the identical physical characteristics has been conveyed previously, test the material for conveyability before purchase and installation.

The air requirements, varying from 4 in. H_2O to 30 in. H_2O pressure and 2–30 cu ft of air per square foot of active porous membrane, are usually furnished by either a fan or low-pressure positive-pressure blower. Nothing fancy is required on these fans or blowers, except for an intake filter-silencer. While the silencer is not required for the fan, the intake filter by all means should be considered. As the air is delivered to the plenum chamber of the conveyor, it is distributed within that chamber and passes up through the porous membrane. Whatever dust or dirt that is allowed to enter the fan or blower intake will invariably wind up in the porous membrane. If sufficient dust and dirt accumulate in the underside of the membrane, its efficiency will be impaired, right down to zero when the membrane becomes plugged. This may not happen for quite some time, but the precaution of initially installing an intake filter will eventually prove worthwhile. It only adds a few dollars to the cost, but can save a considerable amount in maintenance.

This air-activated gravity conveyor has proved its worth in thousands of installations. Its simplicity is unsurpassed; it has no moving parts; it experiences little or no wear; it uses minimum power; it is relatively dustless; and it has great ease of installation, overall operation, and maintenance.

Air Lifts

Air lifts are used for elevating powders and fine grain materials. Heights vary from 15 to 200 ft with capacities up to 350 tons/hr. The heart of the system is the elevator housing (Figure 3.27).

Operation is based on the balance between the material column in the elevator housing and the material flowing upward in the elevating pipeline. A controlled material feed is necessary to maintain the material level in the housing in the upper third so that the pressure from this height of material, together with the conical design of the housing, forces the material in front of the nozzle. The airstream through the nozzle directs the material into the elevating pipeline.

Air requirements (volume and pressure) vary, depending on material being lifted and height of lift.

Packaged Conveyors

The development of the standard packaged conveyor as well as the packaged portable conveying unit began during the early years of World War II. During that time, ports that had unloading equipment to unload grain from ships were being bombed out regularly. To handle the grain in bulk, unload the ship as quickly as possible, and store the equipment in a relatively safe place, the portable pneumatic unit was designed.

For greater mobility, these units were mounted on full highway trailers, built to withstand extremely rugged service, so rugged that today quite a few of them are still in operation in various parts of the world, doing a most remarkable job. They were built to handle grain through a vacuum-pressure

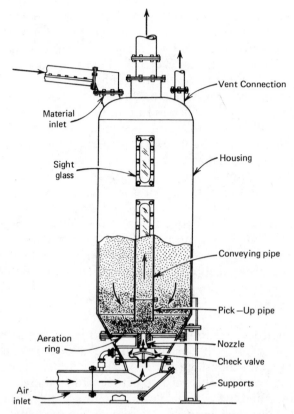

FIGURE 3.27.   Air lift elevator housing. (Courtesy Kennedy Van Saun Corporation, Danville, Pennsylvania.)

arrangement at 40–50 tons/hr, over a combined conveying distance of roughly 250 ft.

In the late 1940s the small portable grain transfer units were developed for use by farmers and small feed and grain mills. From these early beginnings there then came a heavier version of the farm unit, which became the marine type of unit for unloading grain from tankers or other ships to shore points or barges in ports that are void of permanent facilities. These units are of much lighter construction than those built during World War II and, quite naturally, cost less but have much higher maintenance. With these lighter units, the ship's gear, which has a limited lifting capacity, can lift the units onto the deck of the ship. This is an advantage, since in a vacuum–pressure unit using a single blower, the shorter the vacuum leg, the higher the conveying rate. Their accesories are simple, cheap, and expendable. In other words, it is cheaper to furnish a new piece of pipe or hose than to try to repair it.

With the advent of the large explosion in the manufacture and converting of plastics, the portable unit became a common sight. At last count, there were 12–15 different types and makes on the market.

For plastic handling, the units are built with aluminum cyclones, stainless steel or plated rotary feeders, and high-speed blower units; in general they do an admirable job. The small plastic converter can purchase raw plastics in bulk, using the car as a storage bin, and through the use of a small packaged pneumatic conveyor with the wheels removed, since it will stay in one place, he can take advantage of pneumatic conveying and can amortize the equipment within a relatively short period of time.

Most packaged conveyors for the plastic industry and for materials that are relatively free-flowing and nonabrasive are available for activating 2-, 3-, and 4-in. conveying pipelines. Typical conveying rates for handling plastic pellets over a 30-ft suction and a 100-ft pressure distance are 10 tons/hr for a 3-in. pipeline and 20 tons/hr for a 4-in. pipeline. For grain handling, 6-in. units are available in addition to the 3- and 4-in. units; these have a rating of 45 tons/hr over the same 30-ft suction and 100-ft pressure conveying distances.

For plastics processors and molders, many packaged conveyors are available for conveying pelleted and powdered materials from containers or bins and hoppers to machine feed hoppers. Some 20 or more manufacturers use diversity in their conveying methods. Some use straight vacuum; some use dense-phase with small-sized blow tanks; and some use venturis with air pressures as low as 6 psig. Conveying rates as low as 50 lb/hr to match usage rates of the machines being served as well as 5000 lb/hr to serve surge bins or hoppers above the processing machines are available.

Without going into the ramifications of the many packaged conveyors available, the severity of the operation should determine the type of components used for them as well as the other components and accessories.

3.4 AUTOMATED CONTROL

Pneumatic conveyors lend themselves superbly to automatic and automated control. Before we automate, some simple controls and automatics are necessary.

One control needed by both vacuum and pressure systems is the air relief valve. In vacuum systems, this relief valve of the spring-loaded type is installed in the air piping between the receiving station and the exhauster inlet. The purpose of this unit is to bleed air into the air line and the intake to the exhauster so that if the conveying pipeline or receiver station becomes plugged, the exhauster will not be starved for air, which would cause overload in vacuum and power requirement. In pressure systems, although a relief valve between the discharge of the blower or compressor and the feeding unit to the pipeline system may be used, it is much better to use a pressure control switch in low-pressure systems. In the event that the conveying line starts to plug from having too much material enter into it, the pressure will increase. This pressure increase can be controlled by having a pressure control switch interlocked with the feeding mechanism, to stop it at the higher pressure. With no more material going into the line, the line will blow itself out, with the resulting decrease in pressure. After this decrease in pressure, the control switch will reactivate the feeding mechanism, feeding material again into the conveying pipeline. This operation eliminates many plugged lines.

When conveying into storage bins, silos, or other vessels where there is chance of overfilling, high-level indicators should be installed. They should be interlocked with either the exhauster unit on the vacuum system, to shut down that system, or with the feeding mechanism of the pressure system, to eliminate more material being conveyed.

Vacuum and pressure gauges should be installed on the system being used. They should both be equipped with snubbers between the air lines and the gauges to even out the pulsations so inherent with the use of positive-pressure blowers. A good grade of gauge should be used, since it is the only exterior indication of what really is going on inside the system.

When systems become complex, automation and automatic controls become economical. The complex system or systems will convey to or from a multiple number of points and may tie in to weighing and process equipment. When doing this, an electric control cabinet should be readily available to indicate the path the conveying system or stream is following and the operable state of all components and allied equipment.

The panel of the control cabinet should have on its face the flow diagram of the system or systems. Together with this flow diagram, indicating lights should indicate which way diverter valves or pipe switches are directed, components in operation, and condition of material within bins or other vessels to or from which material is being delivered. Together with all these indicating lights should be selector switches, control buttons, start and stop pushbuttons for the operating components, and gauges to indicate operating pressures and vacuums.

With this automatic control, the pipe switches and diverter valves will of course have to be remotely controlled from the panel board. These diverter valves should have limit switches (electrical) so that light indication on the panel board will indicate which direction the switch is pointed. It is better to have the diverter valves and pipe switches equipped with two limit switches so that each switch will indicate that the valve has completed its diversion from one leg to the other. If only one limit switch is used, this could give false indication if the diverter fails to complete its cycle.

A typical control panel (Figure 3.28) used in conjunction with several pneumatic conveyors and process equipment illustrates the ability to operate the systems from a single operating point. On the face of the panel the flow lines of the systems appear. The left-hand part of the panel represents the

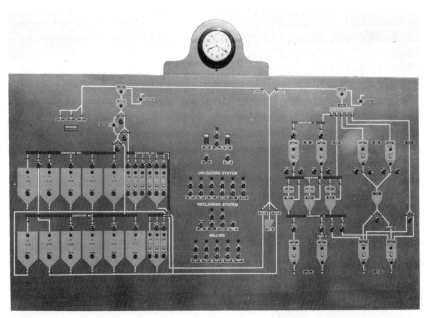

FIGURE 3.28. Typical control panel for pneumatic conveyors. (Courtesy Fuller Company, Bethlehem, Pennsylvania.)

bulk storage section with the flow lines indicating the unloading of materials from the car unloading stations, and delivering to 20 storage bins, and withdrawing from the bins and delivering the materials to the mill room on its way to process. The center portion of the panel contains the main power controls as well as the major controls for each of the unloading, reclaiming, and milling operations. The right-hand part of the panel contains the reclaiming system and the mill room equipment.

Note the ease with which the operator can see how the systems are set up. Lights indicate the path the materials will take; selector switches are placed close by diversion points so that the operator's eyes are focused on both the switch and the indicating lights; when delivering to any bin, the light at the bin will tell him whether or not the bin can take the material he desires to put in it; lights and switches tell him what bin he is withdrawing from, and what mills and scales he is delivering to; also included is a pushbutton to energize each and every light, so that if any light is not glowing when it should either the bulb needs replacement or a malfunction is occurring. Proper interlocking guards against improper operation; warning devices signal when the operation in progress is completed; high-level indicators in the storage bins and scale hoppers are interlocked with the major controls to shut down the system in the event these bins and hoppers become full; all this makes the operation as failure-proof as possible.

In handling bulk materials from bins to processing equipment, the addition of more sophisticated controls can add positive operation to lessen the possibility of human error. Pneumatic conveyors with automatically controlled components can be completely automated to operate in conjunction with other equipment.

The system requiring automated controls is the blow tank. Without them operation is entirely dependent on operator efficiency. The cycle of filling, pressurization, discharge, and depressurization requires automatic inlet and discharge valves, pressure and level sensors, and air supply valves. Their proper sequencing is necessary for accuracy of operation and continuous maintenance of conveying capacities. The design and construction of the control panel are vital to this type of system.

For a description of manual operation, we must turn the calendar back some 40–45 years. A blow tank system was installed to unload bulk cement from a barge and deliver to storage bins on shore. The tank had a capacity of 32 bbl. The cement was fed to the tank by gravity through a hermetically sealed gate operated by an air ram. When the tank was filled, a red light operated by a level indicator inside the tank flashed a signal. After this signal, the operator manipulated two small valves, first to close the inlet gate and then to open the compressed-air line into the tank so that pressure within the tank built up to 22 psig as indicated on a dial gauge. After about three minutes, the pressure began to drop, indicating that most of the cement had been transported out of the tank and that the load in the pipeline was diminishing. When the pressure dropped to 5 psig, the operator closed the air inlet valve. In turn, a bypass vent opened from the tank to the conveying pipeline to clear the line of cement and return the tank to atmospheric pressure to start the next cycle.

Many products of industry have their beginnings in two or more raw materials. These raw materials must be stored in sufficient volume to maintain the production schedules. They must be transported from their storage to metering devices and then to the manufacturing processes. Since pneumatic conveyors can handle only dry bulk materials, we will concern ourselves with just this problem, not liquids. There is no better way to do this than the automatic weighing scale, either batch type or continuous. Scale manufacturers such as Toledo Scale and Howe-Richardson Corporation have developed excellent automatic weighing systems, some with most sophisticated controls. Controls for the scales and the automatic weighing system can be integrated very successfully with the controls for pneumatic conveyors. This integration, together with computer hardware, will give the automation so necessary in large volume handling to assure uniformity of the final product and the ultimate in quality control.

The pneumatic conveyor, when used in conveying materials to weigh hoppers, should receive nearly as much attention as the scale and its mechanism. A sequence of operation must be established so that conveying rates can be set realistically. Time must be allotted not only to the many steps through which an automatic weighing system operates but also to the time required by the pneumatic conveyor to go through its cycle. The first step is the number of batches or weighings required per hour, together with the amount of material required for each batch; next is the time required for the scale to discharge and to stabilize itself between filling and discharging operations. Allowance should also be made for the time required for the controls to act and react. An insufficient allowance for any operation in the weighing cycle can result in choosing a conveying rate for the feeding conveyor that will not satisfy the overall requirement. Likewise, an insufficient allowance in the time required for the pneumatic conveyor to get under way or to clean itself out will materially affect the operation. A little generosity in the time allowances for all these operations will make for a smoother operating system. In addition, the accuracy of a scale and its weighings is only as accurate as its feeding mechanism.

If surge bins are used above the weighing scale, the feeding of the scale becomes a mechanical operation with the pneumatic conveyor used to maintain material in the surge hopper only. If no surge bins are used and the pneumatic conveyor delivers material directly to the scale hopper, certain steps must be taken to assure weighing accuracy.

When handling only one material from storage to a scale hopper and mechanism, the most accuracy in weighing is achieved in the use of a so-called weighing-out scale. This type of scale uses an oversize weigh hopper at least 10% larger than the required batch to be weighed. The hopper is usually filled to a high-level indicator, at which time the level indicator closes the feeding mechanism to the pneumatic conveyor. The pneumatic conveyor is allowed to operate for a short time to clean out the conveying pipeline. To accommodate this additional material, the indicator should be placed in the hopper a sufficient distance from the top or full condition. The scale hopper is discharged through a metering device, which is interlocked with the scale mechanism. When the desired amount of material is discharged, the scale mechanism, whether it be a beam or dial, shuts down the discharge mechanism.

It is desirable to know at all times how much material by weight is in the weigh hopper. The operator should know when starting to discharge a batch that there is sufficient material available to complete the batch. Equipment should be included in the scale mechanism to indicate this. When the weigh hopper is not being discharged, it can be filled by the pneumatic conveyor from the storage bin. The controls for both the pneumatic conveyor and the weighing-out mechanism, including both the scale head and the discharge mechanism, should be interlocked, so that both operations are being done at completely separate times. The filling of the weigh hopper and the weighing and discharging must be two separate operations.

When a considerable amount of material is to be metered and the size of the weigh hopper to accommodate this material would be too large, a smaller weigh hopper and multiple batches must be the answer. In this case, an automatic scale in conjunction with an upper scale hopper is used. The material is discharged into the upper scale hopper continually with the hopper feeding the scale in a weighing-in operation. With batch counters and automatic cutoff equipment for the upper scale hopper and pneumatic conveyor, exact weighing and metering of large amounts of material are accomplished.

When material is discharged directly from the conveying pipeline into the weigh hopper, and exact weight in the hopper is required, the cutoff gate below the storage bin must be closed at some predetermined time before the receipt in the weigh hopper of the correct amount of material. This sometimes is a trial-and-error proposition until the proper controls are correctly set. Once set, variation is very minor.

In conveying into weigh hoppers directly from conveying pipelines, flexible connections must be between the pipeline and the weigh hopper. Furthermore, care must be exercised in the design of the weigh hopper and its supports. External forces to the hopper supporting mechanism must be eliminated to keep these forces from exerting undue effort to the weighing mechanism. These forces will tend to either increase or decrease the force on the scale beam or dial, giving a false weight. Similarly, the force of the incoming air and the pressure of not only the incoming air but also the displaced air in the weigh hopper should be dissipated evenly between the roof of the weigh hopper and the material as it increases in depth. Dust suppression equipment, integral with the weigh hopper, should have minimum air flow resistance. How extensively these points should be explored depends on the accuracy of the weighing equipment and process requirements. If accuracy is extremely critical, design with care.

3.5   FLOW FROM BINS

The flow of material out of storage bins and hoppers can be affected by pneumatic conveyors. It can also affect the pneumatic conveyor used to convey the material away from the bin or hopper. Bin and hopper configurations have received much attention the past 25 years to promote easier and more uniform flow in the discharge of bulk solids. Whereas the size and shape of the bin and its outlet may be accurately designed for gravity flow, the type of feeding mechanism to the pneumatic conveyor must be correctly designed to maintain that gravity flow.

The physical characteristics of the material will have a profound effect on the feeding equipment to the conveyor pipeline. Different arrangements and equipment will be necessary, depending on the type of system used. The combination of the material characteristics and the type of system will determine the feeding equipment and its arrangement.

Considering the material characteristics first, classification is made on their ability to flow. A granular, free-flowing material such as whole grains and uniformly sized plastic pellets requires only a flow-regulating gate to feed the material from the bin outlet to a vacuum system or a rotary feeder to feed a low-pressure system. In the case of feeding into a vacuum system, the negative pressure in the pipeline system will tend to accelerate flow out of the bin, while a pressure system will tend to hinder flow out of a bin by the passage of positive pressure air through the rotary feeder. The granular materials, however, will have sufficient voids between them to allow the air to dissipate up through the material and still maintain flow.

A finely divided material, one that is fluidizable and has the ability to flush, will definitely need a rotary feeder to meter the material into a vacuum system to avoid overloading the system. In this case, the rotary feeder is attached directly to the outlet of the bin. The negative pressure of the vacuum system will also assist in the discharge of material from the bin. To feed this same finely divided material into a low-pressure system, something more than just a rotary feeder is necessary. If the rotary feeder is attached directly to the outlet of the bin, flow out of the bin is often retarded rather than assisted. The positive-pressure air escaping up through the rotary feeder is often greater than the ability of the material to absorb the air and maintain flow. When this condition occurs, air pressure builds up between the inlet of the feeder and the discharging material, causing flow to be reduced and, in many cases, completely stopped. To avoid this situation, some type of mechanical feeder is required to feed the material from the outlet of the bin to the intake of the rotary feeder. Likewise, an air vent either in the mechanical feeder or in a small surge hopper between the mechanical and rotary feeders is necessary to dissipate the air to avoid the buildup of pressure in the outlet of the bin.

In feeding these finely divided materials into fluid solids pumps, the pump and its hopper can be attached directly to the open-and-shut valve at the bin outlet since blowback air from these pumps is minimal, and air binding of the bin is nonexistent. As material is fed into blow tanks and similar equipment, they are usually at atmospheric pressure so the material will flow from the outlet of the bin into the tank without impedence.

In between the granular materials and the finely divided materials, there is a combination of the two. In the feeding of these materials, it is wise to follow the same rules as those for finely divided materials. A good case to illustrate this is an installation put in some 30 years ago handling pebble lime. This lime had a maximum particle size of $\frac{3}{4}$ in. and contained some smaller sized particles and fines. Although the material might be classified as granular, the fines were of sufficient number so that the voids between the larger particles were filled. The engineers, being ignorant of the fact that blowback of air through rotary feeders could air bind bins and restrict flow, attached rotary feeders directly to the outlet flange of the storage bin for feeding at that point, a low-pressure conveying pipeline. The system was a combination pressure–vacuum type, using a single pipeline whereby the vacuum producer at the terminal end of the system operated at 10 in. Hg, and at the intake to the system, a pressure blower, operating at 2 psig, combined to convey the lime over a rather long distance.

For the first few days the system ran smoothly. During this time very little material was in the storage bin. Then material in the bin increased. Flow out of the storage bin into the pipeline system became more erratic and continually lessened until material failed to flow. The increase in material within the bin allowed no dissipation of the air blowback, so the air pressure built up in the outlet and bottom cone of the bin, causing the material to hang there. Note that as long as the air could dissipate up through the material the system operated perfectly, but when it could not, material just would not flow out of the bin.

The only answer to correct the situation was to install screw feeders between the outlet of the bin and the rotary feeders of the conveying system. After this was done, no trouble was encountered. The lesson to be gained here is that when feeding low-pressure systems with partially fine material, mechanical feeders should be used at all times. A pressure as low as 2 psig was sufficient to cause a no-flow condition.

There are many flow aid devices such as air pads, air jets, pulsating panels, knockers, vibrators, and arch breakers available. All of these aids, with the possible exception of air pads and air jets, tend to pack the material in the bin; they generally are not successful if the outlet of the bin is closed while the flow aid device is in action. Whenever these aids are used, they should be interlocked with the feeding and conveying system to give controlled vibration. In the case of low-capacity conveying, it might even be necessary to actuate these units on a momentary basis rather than continuously. This can be done with either a low-vacuum or low-pressure control switch, which would indicate low conveying rate in the conveying pipeline, which could actuate the flow aid until vacuum or pressure returned to its operating level.

In the handling and storage of pulverulent materials, air fluidization assists in the discharge of these materials from bins and hoppers. It also increases the efficiency of the pneumatic conveyor, since the material can be classified as in motion, requiring less inertia to accelerate the material in the conveying pipeline.

Air pads and open-top air-activated gravity conveyors tend to be much more efficient than air jets in promoting flow. Air jets fluidize the material within a small area around the jet, whereas greater air diffusion is achieved with the pads and conveyor. In fluidizing a greater mass of material, the surface tension between the particles themselves as well as between the material and the bin walls and cone is reduced considerably, inducing a more free gravity flow.

4 RESEARCH VERSUS REALITY

Development of pneumatic conveyors through research had its start with Sturtevant's work on dust collectors in 1866; Allington, in 1886, with his work on long-distance, pressure-type pneumatic conveying of fibrous materials; Duckham, in the 1890s, with his invention of the air-sleeve suction nozzle; Mitchell's development of tapering or stepped conveying lines in ship unloading grain in 1906; Kinyon's development of the fluid solids pump in 1919; Gasterstadt, in 1924, with his work on formula development in handling wheat; and Barth, in 1958, who further developed Gasterstadt's findings.

Prior to 1940, most research was conducted with actual installations. During that time industry in general accused those few manufacturers of pneumatic equipment of using them and their systems as guinea pigs. At times, there was quite a bit of validity to the charges. Those of us who lived through the adolescent years (1930s and 1940s) of the industry resorted to what might be termed "kitchen table research." This consisted of visual inspection of a material; hitting materials with a small brass hammer or throwing them against the wall to test for breakage and possible attrition. When we tested abrasiveness, we checked to see if the material would scratch a pane of glass; we tested for bulk density, with a special measuring cup that was filled with material, neatly trimmed, and taken down to the mail room to be weighed on the postal scale. Deducting the tare weight of the measuring cup from the weight on the postal scale, we used a special chart to determine bulk density. Despite its crudeness, this simple research played a very prominent part in the development and acceptance of pneumatic conveying as a viable material-handling technique.

Since the mid-1940s, most research has been done using small pipes, handling at times minute quantities of material. To translate these results into reality (full-sized systems), often the results can be used with no change; at other times, ingenuity is required, restraint necessary, and most of all, the behavioral characteristics of the material being tested must be thoroughly investigated.

One company, after determining optimum conveying conditions (vacuum or pressure, capacity and velocity) in the laboratory, selects a slightly higher velocity for actual installations to assure stable conveying and to accommodate variations in physical characteristics. This is a prudent decision. For its dense-phase, low-velocity-type conveyors, another company prefers full-scale testing, determining exact air–material ratios, velocities, conveying rates, the effects of moisture and abrasion and dust collector requirements. This type of research, however, requires more than just a couple of bags of material.

Normally, when a test of material is to be made, a couple of bags of material are shipped to the laboratory. After receipt, a sample is taken to determine bulk density, fineness (sieve analysis), and moisture content; angle of repose should also be determined. If the angle of repose, after several tests, indicates wide variations in flowability, a look through the microscope may indicate varying particle shapes. If this should occur, velocity and density of the conveying stream will require cogitation after minimum conveying conditions have been determined.

After the test equipment is adjusted for what is thought the best conveying velocity, the bags of material are dumped into the feeding mechanism. In a few seconds the material is swallowed up and is quickly coming out the discharge end. Several more runs will be made, using slightly different velocities, noting pressures used and any unusual happenings.

These tests are over in a short time. Actual systems frequently will run for quite a few hours until they start to plug. What happens is that the velocity is insufficient to maintain the conveying rate day in and day out. It is this type of situation that requires the cogitation so necessary to make the correct decision. Were our optimum conditions sufficient, or should we make necessary corrections to our findings? Despite all the research, the art is still there.

How effective is certain research? In the many studies conducted to determine erosion in not only the pipe bends, but the entire pipeline length, small-diameter pipes (2-in. diameter) short-radius elbows (pipe bends), and mild steel pipe are used. What is used to make the determinations? The answer is that we use a very abrasive material: sand. The industry learned long ago that abrasive materials need something more than mild steel pipes, especially in the medium- and high-velocity systems. They learned that pipe bends had to be reinforced against wear; they learned that rectangular section bends with renewable wearplates were necessary when conveying grain, wood chips, and hog fuel; and they learned that the old term "white iron pipe" was necessary when handling very abrasive materials.

It would be more constructive if instead of mild steel pipe, the bends for testing were made of nickel–steel alloy or other wear-resisting materials. One researcher reported that depth of penetration of the particles into the wall of the bend determines the service life and conveying capacity of the bend. From this conclusion, wouldn't it be more appropriate to use material resistant to wear and give industry some indication of the life we can expect? Or must reality give itself the answer?

About 10 years ago, research was done on the damage to corn from pneumatic conveying. It studied the amount of damage that is caused to pneumatically conveyed corn as a result of the type and condition of the corn conveyed and the operating conditions of the conveying system.

The conveying system was quite elaborate. It consisted of 200 ft of 1.9-in. ID aluminum pipe. Included in this 200 ft were one 45° and fourteen 90° elbows (pipe bends), each having a radius of 24 in. Before going any further, let us ask, "Would any practicing pneumatic conveying engineer design a grain-handling system whose curvature was 22.8% of the total conveying distance of 200 ft?" If he did, his job would be very much in jeopardy.

The blower was sized to deliver a maximum pressure of 10 psig and to deliver air so that velocities ranged from 3960 to 7200 ft/min. If we adhere to the pickup velocity of 55 ft/sec as given in Table 3.10, the maximum or terminal velocity in a system with a pressure of 10 psig would be 92 ft/sec, or 5520 ft/min.

For each test a 7-lb lot of corn was used. How long did it take for this small amount to traverse the system? Possibly 5 or 6 sec at the most. No doubt the airstream was a lightly loaded one. As we all know, a lightly loaded system contributes to the breakage problem. It would have been much better to have used lots of 300–500 lb and loaded the system so that actual practice would have been followed.

After putting the corn through the system once, it was put through the system three and seven times more, for a total of four and eight passes. After the eighth pass, one might have expected cornmeal to be discharged, but kernels were still there. Considering the conditions under which the tests were made, breakage was not too high.

The conclusions reached were that high-moisture (20%) corn was less susceptible to breakage than low-moisture (12%) corn; that velocity should not exceed 5400 ft/min for low-moisture corn; and that if operated improperly, a pneumatic conveyor can cause serious damage to corn.

To prove that breakage can be minimized, a producer of hybrid seed corn in Illinois has been using vacuum-type pneumatic conveyors for close to 40 years. They report they have had excessive breakage at times. However, through proper maintenance, including maintaining butted pipe joints, they presently report little kernel breakage. The systems, designed to operate with minimum velocities, convey at the rate of 300 bu/hr, with a 4-in. conveying pipeline.

When we are required to handle several materials through the same conveying system, how do we determine their compatibility? "Kitchen table research" is a simple answer.

Let us consider the conveying of ferrous sulfate and lime for a water treatment plant. First, the kitchen cabinet should produce three saucers. (Don't use your spouse's good china; you may be in for serious trouble.) One saucer should contain lime only; the second, ferrous sulfate; and the third, a mixture of the two materials. For the correct mixture, consult the treatment plant chemist and request the ratio of the amount of ferrous sulfate to lime that is used to treat a million gallons of water. This same ratio should be used in the mixture for saucer no. 3. Mix the materials thoroughly.

Put the saucers on the window sill to allow the materials and their mixture to be subjected to atmospheric conditions. Depending on those conditions, you will note quite a change. The mixture of ferrous sulfate and lime will look and feel like a material that should come out of a sewage disposal plant. Not only will it be stickey and gooey, but if drying conditions occur, it could wind up as hard as concrete.

In each decade since the mid-1930s, at least one pneumatic conveying system has been installed to handle both materials. Each one had unsolvable problems. The combination of the two materials fouled up filter-receivers, pipe switches, and any other close-fitting equipment. It proves again the statement we've heard so often, "Those who do not remember the past will never inherit the future."

What does this simple research tell us. Simply this: When you handle more than one material in a pneumatic conveying system, make certain they are compatible.

How effective can limited research be? It can be very successful, but it also can lead to unforeseen operating difficulties. It can also prove that Murphy's law—"Anything that can go wrong, will go wrong"—is most applicable.

Some years ago a large company with a central engineering department set out to expand production at a plant 1000 miles away. This expansion called for the receipt and storage of large quantities of Second Clear flour. The flowsheet started with a vacuum-type conveyor to unload railcars and deliver to a transfer station where the flour would be delivered to a pressure-type system for delivery to any one of seven storage bins. From the storage bins, the flour was to be conveyed by vacuum to a screening and batching station; and after screening and weighing, to be pressure conveyed to process.

The first consideration was the type of railcars to use for transportation between the flour mills and the plant. Several bags of two types of Second Clear flour from the same flour mill were sent to the laboratory for testing. After testing, it was decided that Airslide cars would be used. A certain size blower driven by a 15-hp motor was recommended to activate the cars.

During this testing, which was done in early November, it was noted that the bulk density was 36 and 34 lb/cu ft and that the angle of repose was variable and quite steep—above 60°. This high angle of repose should indicate that handling could be sluggish, that bin discharge could be erratic, and that conveying rates could vary.

Samples of these same flours were sent to a screen supplier for testing also. This laboratory reported that the data obtained in its tests indicated conclusively that there should be no difficulty in processing 45,000 lb/hr on its gyratory screen.

It should be noted at this point in the proceedings that only two brands of flour from one milling company were tested. Further, the decision was made also to unload covered hopper cars as well as Airslide and to handle bag shipments out of boxcars.

The central engineering department proceeded with its drawings and the purchasing department sent out its request for bids. Included with the pneumatic conveying equipment were seven glass-lined storage bins with sweep-arm dischargers, the screen, and the weighing equipment, including an electrical control panel for the scaling and reclaiming operation.

The successful bidder presumably based his quotation on the premise that the only flours handled would be those tested. The high angle of repose, however, could indicate material variances, plus the fact that Second Clear flour can vary in ash, bran, and moisture content.

Erection of the equipment was completed the following October and shake-down procedures were initiated. Many problems were encountered. Unloading rates from cars varied considerably; discharge from the storage bins was very uneven; screens in the gyratory sifter became blinded; and vibration was needed to get the flour to flow out of the weigh hopper, even with minimum retention time in the hopper.

Corrections included changing feeder speeds, blower speeds, and screen sizes in the sifter and enlarging the last half of the pressure line to process, to minimize line plugs.

Compounding the picture was the fact that flour was being received from three sources other than the one whose flour was originally tested. With all this going on during severe winter weather, variations in incoming material such as material temperatures, flowability, and sluggishness were very evident. A good example was provided by tests of temperature in the Airslide cars as they were received: One car had a temperature of 51°F at a flour depth of 6 in. and a temperature of 68°F at a depth of 24 in.; a second car had a temperature of 34°F at a depth of 6 in. and a temperature of 48°F at 24 in.; and a third car had a temperature of 70°F throughout the entire car.

Considering the temperature in the second car of 34°F at a depth of 6 in., the temperature at the bottom of the car including slope sheets could be and probably was a lot lower. Just consider the temperature needed to lower the flour temperature to 34°F and have the car travel quite some distance at 30 mph and we have a wind chill factor of at least 0°F. No doubt these temperatures had their effect, especially when activation time for the Airslide cars to start flowing varied from 10 to 50 min. Covered hopper car unloading was more frustrating; material did not flow and had to be air-lanced to the bottom outlets.

The screen supplier, after testing the various flours from the plant, stated that there does appear to be a great variance in the handling and screening characteristics of the flour now being processed. The supplier also stated that on occasion the rated capacity appears to be attained as a result of the particular physical properties of a particular shipment of flour.

Further, the original magnesium alloyed pans of the gyratory sifter had to be replaced with stainless steel. The magnesium alloy pans were severely corroded in irregular patterns. This condition had never been experienced before with several hundred gyratory screens for flour rebolting applications. The corrosion is still a mystery.

A solution was finally reached, but not before making many equipment changes, adding additional screening equipment, and holding many meetings and discussions. How do we avoid such situations, and who was responsible?

First, let us start at the inception of process and equipment requirements. Did Engineering request from Purchasing the sources of supply? Did Purchasing say only one supplier would be used? If yes, then some justification can be made for the procedures used. It is hard to believe, however, that when purchasing the amount of flour with such large usage that a company would tie itself down to a single supplier. Hence, engineering should have been told and should have instigated more diverse testing procedures to include all suppliers of flour. Moreover, we need to know if the company's researchers advised engineering of the differences expected or if they were concerned only with the chemistry of the process. It seems this is the point where materials handling should have been given serious consideration.

Would research in the pneumatic conveying laboratory have been able to pinpoint all the problems encountered? The answer is debatable. It could, however, have eased much of the pain. In retrospect, as much research as possible should have been done on all flours, not just two, plus a little "kitchen table" research thrown in.

This "kitchen table" research would consist of a refrigerator with a freezer compartment, an oven, a cookie sheet, a thermometer, and a 1-gallon container with a lid. In the lid, a vent hole should be provided, together with a hole in the center for insertion of a thermometer into the material within the container.

At ambient temperature, we can make the usual tests, such as aeration characteristics, angle of repose, bulk density, and conveyability in the laboratory. We should also determine the effects of temperature changes. By filling the metal container about seven-eighths full, we can cool the material in the refrigerator, freeze it in the freezer compartment, or heat it in the oven to any desired temperature. Visual inspection of the material under any of these temperatures could give us an insight into what, in reality, we may expect. When determining the angle of repose at these temperatures with the cookie sheet, it should have the same temperature as the container.

Finally, who was responsible? Was it the vendors, or was it the purchaser and his engineers? You be the judge.

When we all know what we should about the material to be handled, how do we go about designing the system and the plant layout? Do we take the advice of Albert Einstein when he said, "Everything should be made simple, but not simpler," or do we complicate things with a grandiose system plus an elaborate electric control panel that lights up like a Christmas tree?

With computer control and intricate production processes, many times the complicated system is necessary to allow automation to exert complete control of the handling and manufacturing process, aiding cost reduction and product uniformity and avoiding human failure. What do we do, however, with what should be a simple system to unload cars of material and deposit it in closely grouped storage silos? Do we use a straight vacuum system or do we use a vacuum system for car unloading and then a pressure system for distribution to the bins?

Let us consider a malt unloading system. The first consideration should be that we handle the malt in such a manner that breakage of the kernels and skinning of the hulls are kept to a minimum. (When handling chemicals or other materials where attrition or particle breakage is inconsequential, this perrogative should be disregarded.) Then we must consider the mode by which the malt is received. The plant is too far from the source of supply for truck delivery, so the only method is railcar. Checking with the suppliers and railroads, we find that 100-ton-capacity covered hoppers are available and in sufficient numbers to avoid a shortage problem. Then we find that the cars available have three discharge hoppers equipped with built-in 6-in. nozzles for pneumatic unloading. If, however, and despite assurances, covered hoppers are not available and boxcars must be substituted, equipment should also be on hand to unload these cars.

Two equipment arrangements are illustrated for comparative reasoning. Figure 4.1 illustrates a system to unload cars from four unloading stations by vacuum, delivering to a transfer station consisting of a filter-receiver, malt sampler, magnetic separator, 2000-lb garner scale with upper- and lower-scale hoppers, and the necessary exhauster unit. From the discharge of the lower-scale hopper, a pressure system is used to transfer malt to any one of five storage silos consisting of a blower unit, rotary feeder to feed malt into the pressure-conveying pipeline, four remote-controlled fully enclosed pipe switches with 30° angle off-legs, and piping connections to the silos. Since we are blowing the malt into the silos, we must provide equipment to alleviate the air. This is done by a separate dust collecting system, including a cloth-tube dust collector, piping from each silo to a main header, and a fan for actuating the system. Figure 4.2 illustrates a straight vacuum system for unloading the cars and delivering to the top of the storage silos

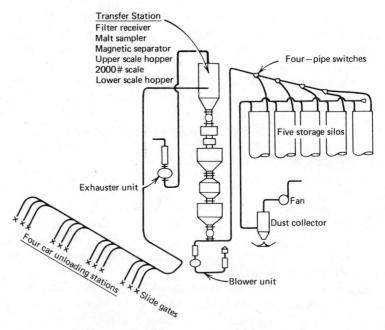

FIGURE 4.1.   Vacuum-pressure unloading system.

consisting of a filter-receiver, 6-bu-capacity automatic grain-receiving scale with upper- and lower-scale hoppers, and a screw conveyor for distribution to the silos. The magnetic separator and malt sampler can be installed in the lower-scale hopper at its discharge into the screw conveyor. This appears to be a more desirable location for the sampler, since it will not only take

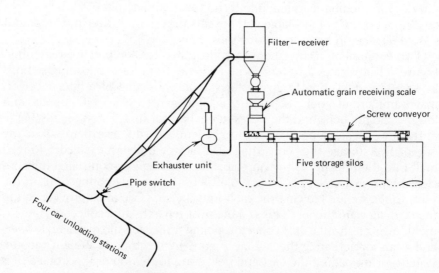

FIGURE 4.2.   Vacuum unloading system.

the necessary samples but also indicate the condition of the malt in the sampler bucket to indicate whether or not excessive breakage or skinning of the kernels is occurring.

Some engineers consider it advantageous to install a multiplicity of unloading pipes for connection between the car hopper outlets and the conveying piping. They aver that minimum length of hose is required and that installation of slide gates in the pipe just above the hose connection will allow the operator to attach the hose to the next hopper while the first is unloading, thereby reducing time loss when going from one hopper to the next. It is doubtful that much time is saved. The average distance between hopper outlets is 12 ft. By using a simplified arrangement as shown in Figure 4.2, very little time is used to disconnect the hose from the car hopper, walk 12 ft to the next, insert the hose, and resume conveying. An additional 12–15-ft length of hose will be required. This additional length of hose will cause much less damage to the malt than the many laterals it must go through in Figure 4.1.

Let us list and compare the prime equipment requirements of the two arrangements:

Equipment	Vacuum	Vacuum-Pressure
Exhauster unit	100 hp	100 hp
Filter-receiver	One	One
Rotary feeder and drive	$1\frac{1}{2}$ hp	$1\frac{1}{2}$ hp
Upper-scale hopper	16 cf cap.	100 cf cap.
Automatic scale	6 bu	2000 lb
Lower-scale hopper	16 cf cap.	100 cf cap.
Unloading piping system	7 bends	16 bends
	2 laterals	11 laterals
	1 switch	12 slide gates
Blower unit		50 hp
Rotary feeder and drive		$1\frac{1}{2}$ hp
Pressure piping system		9 bends
		4 pipe switches
Dust collector		One
Dust collector fan		15 hp
Distributing screw conveyor	One	
Screw conveyor drive	$7\frac{1}{2}$ hp	
Bin valves, motor operated	4	

It is evident from this comparison that the vacuum system requires less equipment than the vacuum-pressure. This should result in a less costly system, not only for equipment but for erection and building costs as well. Electrical equipment also will be less, including a much simplified electric control panel.

The principal objection to the vacuum system will be the location of most of the equipment atop the storage silos and the resultant tall structure. This requires, however, only minimal housing compared to what is needed to house the larger-scale equipment, dust collector, and three air-activation units. Operating power costs will also be less. At a cost of 5¢/kW hr, operating savings on a 40-hr/week operation can approach $5000 per year. Last, but not least, malt breakage in the vacuum system will be but a fraction of that in the vacuum–pressure system, which in reality was considerable.

5 | APPLICATION TO INDUSTRIES

Pneumatic conveyors applied to the handling of materials between segments of manufacturing processes have provided adaptability and complete reliability, thus affording economies in not only the process but also the overall plant layout and arrangement. Plants whose process segments had to be closely coupled because of a material-handling problem are now able to separate those segments for improved plant operation. Where segments must be separated, the pneumatic conveyor gives a positive, relatively inexpensive method in transporting the material from and to the segments involved. The manufacturing facilities and the storage and shipping silos are often a considerable distance apart, and the material is transported most effectively through a single pipeline.

When new plant facilities are designed, the handling of materials—incoming, in process, and shipping—is frequently not given the consideration it deserves. Materials handling is only considered after all other equipment has been located. This makes the materials-handling engineer's job most difficult, since he is obligated to move materials through circuits that are very complicated and that could be simplified if the proper consideration were given in the beginning. The materials-handling engineer should be a vital member of the designing team, exerting influence on the overall design and plant layout. It is with this thought in mind that the applications of pneumatic conveyors to various industries are discussed so that they may be an integral part of the whole plant operation.

5.1 CEMENT INDUSTRY

The cement industry, from manufacturing through shipping to distribution and use, has received the greatest impact of any industry from the use of pneumatic conveyors. It was this industry that accepted the fluid-solids pump as its material-handling workhorse and that spawned the air-activated gravity conveyor. It also gave high usage to the blow tank in its conveying operations.

The cement mill was probably the first basic industry to include in its manufacturing circuits pneumatic conveyors as standard conveying practice in the handling of the material between the various stages of manufacture. A typical flow diagram (Figure 5.1) indicates the prime equipment and the applicable conveyors in the manufacturing circuit. We shall refer to the numbers of the figure to pinpoint where pneumatic conveyors are used.

Cement has its beginning in limestone rock and other materials containing lime, plus other materials of proper chemical properties for conversion in a burning process, usually a rotary kiln. These raw materials are much too coarse for handling in pneumatic conveyors. The materials at this stage of manufacture are all handled by mechanical conveyors through the primary and secondary crushers, and to the raw grinding mills (1 to 5). The first application of pneumatic conveyors is in the discharge of the raw grinding mill (5). This is an air-activated gravity conveyor, not exactly a pneumatic conveyor by definition, but a member of the family as stated in Section 2.9. An air-activated gravity conveyor (6) is ideally suited to convey horizontally the discharge from either one or more mills to bucket elevators, which in turn discharge into the air separator (7), for separating the fines and coarse material. After this milling and separating operation, the fine cement raw material is then ready to pass on to what might be termed the conversion part of the manufacturing process.

At this point, the material is finely divided and easily conveyed by a fluid solids pump (9). The pump system discharges either into the kiln feed tank directly or into blending silos if additional blending of the raw materials is required for the proper burning blend. If blending silos are used, the discharge from these silos should be closely controlled for delivery of the blended material to either an air-activated gravity conveyor or a fluid solids pump (15), for delivery to the kiln feed tank.

From this point, the cement raw material is delivered to a preheater to preheat the material prior to burning and to recuperate heat from the exhaust gases from the kiln. After the raw material passes through the preheater, it is burned in the rotary kiln and discharged from the kiln as cement clinker. The clinker usually is cooled in an air-quenching cooler next, then passed through a clinker breaker and transported to the finish grinding mills. Between the clinker breaker discharge and the intake to the finish grinding mills, the clinker is too coarse for handling in pneumatic conveyors, so mechanical conveyors must be used.

From the discharge of the finish grinding mills (25), as in the case of the raw grinding mills, air-activated gravity conveyors (26) are used to convey from one or more mills to a fluid solids pumping system (29), which in turn conveys the material to the cement storage silos. These silos now store the

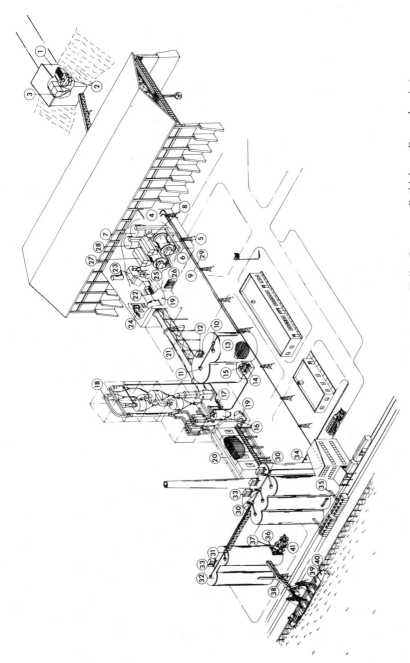

FIGURE 5.1. Flow diagram, dry cement mill. (Courtesy Fuller Company, Bethlehem, Pennsylvania.)

143

finished product, and from which the cement is delivered to packhouse, railcars, trucks, barges, or ships.

For loading trucks, air-activated gravity conveyors are used to convey the cement from the outlets of the silos to a single delivery point over a truck weighing scale, where the trucks are filled to a predetermined weight. In the same manner, covered hopper cars are loaded either by gravity spouts if the track is close enough to the storage silos to satisfy gravity spouting or, if the track is a short distance from the silos, by swivel air-activated gravity conveyors (36). In the loading of barges, either the air-activated gravity conveyor mounted on a swivel boom (38) or the fluid solids pump is used, depending upon distances involved. For loading ships, either the air-activated gravity conveyor or the fluid solids pump, or a combination of the two, is used.

For flash calcining and coal firing of the rotary kiln, a stationary flash furnace is installed between the outlet of the preheater and the feed inlet to the rotary kiln. Coal is ground in a grinding mill. The resultant pulverized coal is delivered to a surge bin, which in turn feeds two fluid solids pumps (Figure 5.2). One pump delivers the pulverized coal through a single pipeline to a four-way stream splitter, which feeds the coal to four burner points in the calciner. The other pump delivers the pulverized coal to the burner hood of the rotary kiln. Through the use of fluid solids pumps, a pulsation-free delivery of coal mixed with the conveying air assures a constant burner flame.

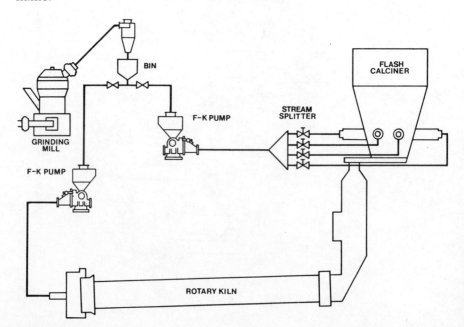

FIGURE 5.2. Pulverized fuel feed for coal-fired kiln. (Courtesy Fuller Company, Bethlehem, Pennsylvania.)

As we can see, pneumatic conveyors are used quite extensively in the manufacture of Portland cement, permitting flexibility of plant layout. They have eliminated considerable dust from the manufacturing process, reduced maintenance materially, and made the cement mill a much safer place in which to work. The alternate, as we can well imagine, would be a maze of screw conveyors, bucket elevators, and possibly belt conveyors, all of which would require much more maintenance, and possibly an overall higher manufacturing cost.

In new mills, where computers control the blending and burning process of the raw material, continuous sampling and x-ray analyses for proper blending of the raw materials are necessary. To bring the samples from the kiln feed and raw mills of the process to the laboratory and computer operation, small pneumatic conveyors of the vacuum–pressure type enable the samples to be delivered without manual collection. This further enables the mill operators to know at all times the quality of the blend of raw materials and to make prompt corrections.

The cement, after being manufactured, must now be distributed. The distribution through the use of air and pneumatic conveyors has developed so much within the past 50 years that the 1-cu-ft bag of cement has dwindled from a commonplace item to a rarity. Bulk transportation took a great step with the development of the covered hopper railroad car. The development of this car as well as the development of the fluid solids pump and blow tank for unloading the car went hand in hand. The very first covered hopper car to be loaded with cement was loaded by a Fuller-Kinyon fluid solids pump at the Hercules Cement Company's plant at Stockertown, Pennsylvania, around 1924. With a method for conveying the cement from the outlet of the car to the user's storage bins that was relatively dustless and maintenance free, the use and shipment of cement in bulk has increased until today possibly as much as 90% of the cement being used is handled in bulk.

With covered hopper car distribution, the use of bulk cement was limited to plants having railroad sidings. With highway building on the rise after World War II, as well as many large construction projects some distance away from rail lines, additional methods of distribution were urgently needed. This requirement brought forth the pneumatic truck trailers that use either the pressure differential method or the Airslide pump method for unloading the trucks and conveying the cement through a pipeline into the storage bin. Economies are also afforded the small ready-mix plant, off rail facilities, which can take advantage of bulk purchases, so that it is competitive with those having rail facilities. In fact, today much short-haul cement distribution is done by the pneumatic truck, rather than rail. While transportation costs vary, it is more economical to ship by truck than by rail for distances up to 250 miles. Above 250 miles, rail transportation is cheaper.

The ready-mix plants and large construction concrete mixing plants make good use of the fluid solids pump and blow tank for unloading the cement from cars and delivering to their storage silos. Such installations give the users reliable unloading equipment and economics of bulk shipments.

With competition increasing, the costs of distribution and delivery have come under close scrutiny in order to make the cost to the user at his plant as low as possible. From these studies came the distribution stations at various points on our waterways, which afforded a wider and broader market. These distribution points are supplied with cement by either barge or ship delivery from the mill. From these distribution stations, the cement is loaded into either railcars or trucks for delivery to the customers' plants.

Air has played a most important part in the development of marine distribution of bulk cement. Marine transportation includes barges and ships from as small as 1800-barrel capacity to as large as 90,000-barrel capacity (1 barrel = 4 cu ft). Most of them are self-unloading, being equipped with air-activated gravity conveyors and Fuller-Kinyon fluid solids pumps to convey the cement from the holds of the barge or ship to shore storage facilities. There are no standard designs because each vessel design creates its own combination of conveying machinery. Unloading rates can vary from 150 barrels/hr to 4000 barrels for each installation. Large ships in the range of 60,000 to 80,000 barrels of cement may have as many as four pumping systems, allowing an unloading rate in excess of 4000 barrels/hr system.

At the customers' plants, whether it be a large construction site, a ready-mix plant, or a cement block and pipe manufacturing plant, storage facilities must be available to store the bulk cement. The size of these storage facilities must be based on consumption, rate of receipts, and sufficient storage volume to maintain production between receipts. When rail delivery is used, conveying equipment is necessary to get the cement out of the covered hopper car and into the storage bin.

This conveying operation is a prime application for the fluid solids pump, air-activated gravity conveyor, and the blow tank. Depending on the location of the track in relation to the storage bin, together with the physical terrain, any of these can be used to advantage. If the conveying distance from the car outlet to the inlet of the storage bin is greater than 125 ft, the fluid solids pump or the blow tank should be used.

In using the fluid solids pump or the blow tank, if the discharge from the car is to go directly to the inlet of the pump or tank, it must be installed directly below the track, which means a watertight pit with concrete walls and a supporting structure for the railroad track. Although a pit will still be necessary in most cases, it can be installed alongside the track, eliminating the structural support for the railroad track. A method of conveying the cement from the car to the pump hopper or blow tank must, however, be used. Mechanical conveyors can be used, but a much more effective conveyor would be the air-activated gravity type, even though additional height would be necessary to maintain the slight slope that is necessary. This would deliver the cement to the conveying system in a more fluid state, which would make the conveying system more efficient. It would also keep maintenance costs at a minimum.

In this unloading arrangement, we can go a few steps farther and consider putting the storage bin adjacent to the track, even though the use point might be 200 ft away. This would call for an operation that is required not only to unload cars but also to deliver the cement to the use point. With the storage bin adjacent to the track, a single pump, located directly below the storage bin, can be used for both operations, although not simultaneously. The operation can, through simple automatic controls, maintain material at the point of use and still unload the cars. Such an installation would use an air-activated gravity conveyor to convey from the car to the pump. The pump hopper would accept material from both this air-activated gravity conveyor and the outlet of the storage bin. An automatically operated two-way diverter valve in the conveying pipeline will divert the conveying stream to either the storage bin or use point.

When the conveying distance from the track to the inlet of the storage bin is below 125 ft, the Fuller-Kompact fluid solids pump, installed in a shallow pit below the track, is an economical system for unloading the incoming cement. This is advantageous not only to the concrete-batching operator but also to the asphalt-mixing plant. Pulverized limestone or filler dust for the asphalt mix can be unloaded by this same pump. Illustrative of the savings involved is a highway construction plant requiring the receipt of 600 barrels of cement per day for the concrete-batching plant, and limestone in varying amounts for the asphalt plant. Prior to the installation of the pump, to satisfy the cement requirements for two days, 12 truckloads were received, with each truck on the road 4 hr and each taking 30 min for unloading. Limestone for the asphalt plant was received in 50-ton boxcars of 100-lb bags, which required $12\frac{1}{2}$ manhours to unload. After the installation of the pump, cement was received in 400-barrel capacity covered hopper railcars, which were unloaded in 75 min. The limestone also was received in covered hopper railcars and unloaded in 75 min. By receiving the limestone in bulk, an additional saving of $1.00/ton was effected.

For the smaller plant, bulk receipt by the self-unloading truck is more economical. The only equipment required by the plant is a 4-in. rubber hose and pipeline to connect the discharge pipe of the truck to the inlet of the storage bin, and a dust collector on top of the storage bin to satisfy the air delivery of 400 cfm, which most truck power plants employ. Trucks should, however, be able to drive right up to the storage silo to maintain the shortest conveying distance for maximum rate of unloading.

Further evolution in the distribution of cement is the self-unloading car of the pressure differential type. They do not carry their own air supply, like the self-unloading truck, but a permanent air supply station at the unloading site is all that is necessary to activate the car.

5.2   BAKING INDUSTRY

The application and use of pneumatic conveyors in the baking industry had its greatest expansion with the introduction of the specialty railroad cars for the transportation of bulk flour and other materials requiring the utmost protection during transit. Through World War II and the late 1940's, all flour was shipped in bags in boxcars. Stricter sanitation requirements together with the need for reduced packaging and distribution costs necessitated bulk flour cars. Sanitation alone in shipping flour in bags was costly and troublesome when we consider the typical markings on boxcars such as "Special Weevil Control Car, Do Not Contaminate, Return Empty to Buffalo for Flour Reloading." In the flour mill, the bakery, and the package food plant, pneumatic conveyors have, to a great extent, replaced the mechanical conveyor. The reasons for this changeover are principally the sanitary and self-cleaning features of the pneumatic system.

The year of 1948 is a commemorative one for the flour milling and baking industries. In that year, during the season when the natural landscape changed from lush green to the browns and reds, the landscape of the modern bakery began its change from warehouse storage of bagged flour to the vertical and horizontal storage of bulk flour in bins and silos. To Fuller Company, Bethlehem, Pennsylvania, can go the lion's share of credit for making this transition a reality.

Experimentation in the handling of foodstuffs in finished form was begun by Fuller Company during the early part of 1946. The pneumatic conveying of flour, while requiring special care, presented no insurmountable problems. The real problem was to find and develop a suitable railroad car for the transportation of flour in bulk from the the flour mill to the bakery. In conjunction with two railroad car-building companies, two distinctively different types of car were developed. One of these, the General American Transporation Corporation's Trans-Flo* car, gained the most favor. It was this car that gave the impetus to the design of bakeries whose entire flour and sugar handling was done by pneumatic conveyors. The impetus, however, had several obstacles to overcome.

The bakers, many of whom insisted on aging their flour for several weeks before usage, were reluctant to go to bulk storage. They also did considerable shopping around for extremely small price differentials for their bagged flour. This shopping around would be greatly reduced since they had to deal with few suppliers who, at the time, could ship in bulk, and who would be less prone to reduce prices. When bulk was considered, however, bakeries found their handling costs were reduced so much that the price differentials were of minor consequence. Even after learning the great reduction in handling costs, pneumatic conveyors had to be proved to be more sanitary,

---

* Registered trademark, General American Transportation Corporation, Chicago, Illinois.

free from fire and explosion hazards, and to aid, not impair, the quality of the product.

The proof of greater sanitation was plainly evident in that the pneumatic conveyor is self-cleaning. With properly designed equipment where square corners and joints were filleted, areas of buildup were eliminated, obviating the need for frequent cleaning. Further, the pneumatic conveyor would reduce the propogation of infestation.

The reduction in the fire and explosion hazard was conclusively proven to the insurance authorities. Through tests conducted in 1947, proving that pneumatic conveyors do not propogate local combustion, their use was approved, as was the use of bulk storage bins. The bulk storage bins must be designed to include explosion vents, whose size varies in relation to the total volume of the bin. There appears to be no strict standard between insurance companies on the ratio of vent area to bin volume, so the designer should check with the insurance company involved to get the approved ratio.

After the obstacles were removed, every large bakery, either newly constructed or modernized, went to bulk handling and storage. Some have vertical, and others have horizontal storage bins; they all use pneumatic conveyors for their unloading and conveying operation; and they all use a great deal of automation. What prompted this diversion from sack to bulk?

First, railcar improvement through the development of General American Transportation Corporation's Airslide car. This car required only one connection between it and the pneumatic conveying system, whereas the Trans-Flo car required six such connections and required much more attention and interrupted unloading rate when transferring from one hopper outlet to the next. By using a number of leased cars to satisfy the bakery consumption, together with the required inventory in storage, and pneumatic conveyor unloading of cars, a saving of 25¢/cwt (sack) was achieved (80¢ in 1982). This saving alone gave fruitful thought to bulk handling.

With the flour now in bulk storage, shrinkage due to broken sacks, residue remaining in empty sacks, disposal of empty sacks, and sanitation labor is eliminated. By using a second pneumatic conveyor to convey from the bulk storage to the weighing and metering station for feeding the dough mixers, further economies are available, as well as greater and more uniform quality control. Labor is reduced within the plant, reducing manufacturing costs. Automatic weighing and batching makes a more uniform product so necessary for continued and increased sale of the ultimate product.

Where quantities such as those required in large bakeries must be handled, several pneumatic conveyors are employed. Such an installation is shown in Figure 5.3. Although this flour and sugar handling system was built around 1951, it still is considered to be a prime example of the application of pneumatic conveyors and the automation under which they are controlled. Usage of flour, sugar, and cerelose dictated that two independent

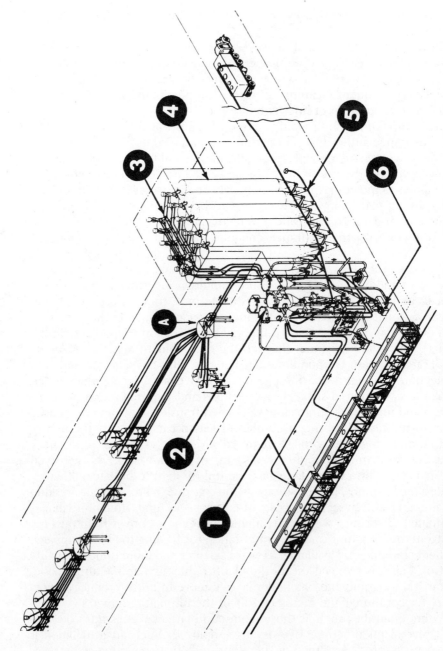

FIGURE 5.3. Pneumatic conveyors handling flour and sugar. (Courtesy Fuller Company, Bethlehem, Pennsylvania.)

150

pneumatic conveying systems would be required to unload these materials from three car positions and deliver them to any of 20 storage silos at a rate for each system of 20 tons/hr. Bulk sugar was also to be unloaded from a bulk truck and deliver to bins. The system also was to reclaim flour or sugar from the storage bins and deliver to any 9 process bins at the same rate of 20 tons/hr.

In the figure, the bulk flour cars (1) are spotted on the receiving siding adjacent to the conveying pipelines. Before unloading begins, samples of flour are taken from the car to the laboratory for analysis. If a positive analysis is received, the car unloading can begin. The operator attaches the flexible car hose to the outlet nozzles of the flour cars, and when Airslide cars are to be unloaded, he attaches the air supply hose to the car's plenum chamber piping. He then sets the proper switches from push buttons and selector switches on the electric control panel to direct the flour to the desired storage bin. Next he presses a button, and through sequence starting of the various pieces of equipment that make up the system, the system is activated. With the system operating, the air for the Airslide car is then supplied and flour empties out of the car, through the system, and into storage.

From the car, the flour is conveyed by the vacuum leg of one of the systems to a filter-receiver (2) where the flour is separated from the airstream and discharged through a rotary airlock. From the airlock outlet, the flour drops by gravity into a stream splitter for passage through a bank of Entoleters to a rotary feeder. This feeder feeds the flour into the pressure leg of the system, which is activated by the same exhauster-blower (6) that is used to activate the vacuum leg. The flour is then pressure conveyed to its proper storage silo through the pipeline, and motorized pipe switches, to a cyclonic separator (3). This cyclonic separator serves two silos (4), so a remote-controlled electric motor diverter valve located at its outlet directs the flow to the proper silo. The conveying air is returned to a dust return filter-receiver, which separates the flour dust that passed through the cyclone separator from the airstream. This flour dust is returned to the pressure leg of the system for returning to the storage silo.

During this unloading operation, the operator is free to refill the use bins as required. Before a particular run, the mixing foreman sets the hopper scale below the use bins to scale successively the right weights of flour, sugar, salt, and dry milk into the mixer on the floor below. Panel-mounted high- and low-level indicators for the use bins guide the operator in maintaining a supply of material in those bins. When a low level is indicated, the operator sets the proper switches and conveying line circuit from the panelboard. He then presses a button, as in unloading, to activate the system. When the use bin is full, a high-level indicator gives an alarm, and shuts down the feeding mechanism below the storage silo (5). Through a time-delay relay, the system is allowed to run for a short period of time to clean itself out.

This installation vividly illustrates the operating flexibility that can be attained through the use of pneumatic conveyors and pipeline conveying. Unloading is possible from one railroad car or one highway truck at a time, or from two railroad cars or one car and one highway truck simultaneously. Delivery of materials to the silos or use bins can be made in a variety of combinations, all set up by controls from the panelboard. From a single unloading point, material can be delivered to either a storage silo or a use bin. From the two unloading points, material can be delivered to either of two storge bins simultaneously. From one unloading point, material can be delivered to a storage silo, and from the other unloading point, material can be delivered to a use bin, also simultaneously. From a storage silo, material can be delivered to any one of nine use bins, however, only one system at any time can be used to deliver material to a use bin. Finally, material can be delivered from a storage silo to a use bin, and from a single unloading point to a storage bin simultaneously.

The master control board contains 162 indicating lights, 50 push buttons, and 5 selector switches, all mounted on a schematic diagram of the entire conveying operation. As control settings are made, the indicator lights flash on to show the exact course of the material through the system. The operator can see at a glance the point from which material is being conveyed, the route over which it is traveling, and the point to which it is being delivered. Once the correct setup is made, the entire operation is automatic. Since the lights are very important, they can all be turned on at once for test purposes. In the overall picture, the pneumatic conveying of all bulk materials has been an important factor in bringing the total handling costs down to approximately 5% of total manufacturing costs.

Pneumatic conveying and bulk storage can reduce costs for not only the large bakery but the small bakery too. Containerization in the form of Tote* Bins, sanitarily constructed of aluminum, and the U.S. Sealdbulk system of collapsible containers of reinforced rubber construction have brought bulk to off-rail and small user.

The Tote Bin is a weather-tight, hermetically sealed container, commonly made of 6061T6 aluminum alloy. They are offered in standard sizes of 42, 74, 90, 98, and 110 cu ft capacity. Intermediate sizes to suit individual needs or specific problems are also available. The standard sizes will hold from 1500 to 3850 lb of flour. Standing on legs $4\frac{1}{2}$ in. high, they can be moved easily with a fork or pallet lift truck. A pivoted hopper tilt which raises and tilts the Tote for emptying is available. Usually the screw discharge type is used when pneumatic conveying is employed to convey to either a use bin or directly into a scale hopper. The discharge from the screw is directed to the inlet of a rotary feeder for charging the flour into the pipeline of the pneumatic system. If sufficient height is available, the screw conveyor

---

* Registered trademark, Tote System, Beatrice, Nebraska.

type of discharge might be replaced by an air-activated gravity conveyor, also terminating at the inlet of the rotary feeder.

The U.S. Sealdbulk system, by Uniroyal, uses the Sealdbin Collapsible Container. Sealdbins are built like a tire, with two or more plies of high-strength cord fabric and synthetic rubber vulcanized into one piece. Sizes available include, 70, 300, and 370 cu ft. They will hold 2450, 10,500, and 12,950 lb of flour, respectively. Emptying of the Sealdbin can be directly to the pneumatic conveyor by connecting its unloading sleeve to the intake of the rotary feeder. In a low-pressure system, however, blowback from the rotary feeder may cause an intermittent discharge. The Sealdbin, during emptying, is supported on an unloading cradle. The hydraulic arms of the cradle raise gradually to funnel the flour to the emptying closure, which includes the rubberized nylon sleeve. This sleeve also may serve as a pinch valve to control flow if necessary.

If bagged or sacked flour is received, a pneumatic conveyor still may be applicable and most advantageous. As the sacks are received, instead of being stored as sacks in palletized piles, they can be dumped into a pneumatic system for storage in bulk in horizontal bins. This eliminates a good deal of housekeeping, eliminates spillage from broken sacks, minimizes the area required for comparable storage volume, and can reduce in-plant labor by eliminating rehandling of the sacked flour. A dustless operation can result by use of a bag dump hopper equipped with an air-activated gravity conveyor to feed the rotary feeder and a sack cleaning nozzle connected to the pneumatic conveyor's dust collector in the hood of the dump hopper. This nozzle will reduce the residue remaining in empty sacks after being dumped. Assuming that 10¢ worth of flour is reclaimed from each sack, and the usage is 1000 sacks/week, the saving would be $100/week, or $5000/year.

The medium-sized bakery can take advantage of railcar delivery and bulk storage through the use of horizontal storage bins. These bins can be installed in existing buildings with little or no alterations to the structure. By using an air-activated gravity conveyor for discharging the flour, the bins are self-cleaning. Conveying pneumatically to the scale hopper is simple, effective, and efficient. Quick-acting scaling valves in the conveying pipeline allow accurate weighings in the scale hopper. When the proper weight is reached in the scale hopper, the weighing mechanism activates the valve, returning the flour in the conveying pipeline to the storage bin, and simultaneously stops the feeder at the storage bin so that no additional flour is charged into the system.

The pneumatic conveyor effected another advance in the flour milling industry. This advance became a reality in 1949, even though a patent was issued 40 years earlier. The first words of the German Patent Specification No. 256651 issued July 29, 1909, held by the firm of Greffenius, were: "The present invention concerns a milling plant in which conveying of all products is exclusively carried out pneumatically from a central point." It took the development of the high-efficiency cyclone to render the far-reaching invention effective in practice. Today, in conveying stocks during the various stages of milling, fans activate as many as six separate pipelines. Careful regulation of the air to each pipeline results in effective and reliable operation. It is difficult to believe that any milling company considering a new plant would seriously think of mechanical handling as a challenge to pneumatic methods and the all-pneumatic mill.

5.3   BREWING AND DISTILLING INDUSTRY

With the repeal of the Eighteenth Amendment to the Constitution of the United States of America in 1933, the old method of shipping and receiving grain in bags to both the brewing and distilling industries was found to be outmoded. The reactivation of the legal manufacture of beer and distilled spirits and the insatiable thirst of the populace caused these industries to look for and take advantage of the most expeditious manner of handling their raw materials. The vacuum type of pneumatic conveying system allowed them to receive their grain and adjunct materials in bulk, in boxcars, reducting their cost, making the unloading of cars easier and materially improving their in-plant operations.

Large and small brewers and distillers took immediate advantage of bulk handling, which had much to do with the advance and more univeral use of the pneumatic conveyor. In the brewing industry, it normally takes one bushel of malt and 19 lb of corn grits to make one barrel (31 gallons) of beer. The small brewer who made 100,000 barrels of beer per year found that he had a daily usage of 400 bushels of malt and 7600 lb of grits. In a week's time, he found that he was using slightly more than a carload of the two materials. By using the rule of thumb that the minimum receiving requirement of one carload per week was necessary to adequately amortize his capital investment, he found a ready answer to the then new handling technique, bulk storage and pneumatic conveying. While this might be an isolated case, it is typical of what can be done. A small brewer in eastern Pennsylvania has received such dividends from his bulk installation since 1943 that today he is able to defy the bigness now prevalent in the industry.

This brewery received malt and grits on a railroad siding that was on the opposite side of the brewery from the brewhouse and existing wooden storage bins inside the building. Although bulk storage bins were used for years, the malt was received in boxcars, in bags, as was the grits, and unloaded from the cars by a six-man labor force. The bags of material were loaded onto a truck, which then was driven around to the other side of the brewery. At this point, the same six-man crew unloaded the truck, placing the bagged material on the freight elevator and taking it upstairs to the top of the storage bins. At the top of the bins, the bags were transported some 40 to 50 ft and then dumped into the bins. The empty bags, which could be returned for a rebate, were piled and tied up, requiring possibly more manhours than what they were really worth. A truck and 48 manhours of labor were required to unload the single car of material. After the installation of a single vacuum type pneumatic conveying system to unload malt and grits from the boxcars at the rate of 450 bushels of malt, or 12,000 pounds of grits, per hour, to deposit the malt and grits directly into the storage bins and to be able to reclaim the malt and grits from the storage bins to the weigh hoppers in the brewhouse, the installation was amortized within the short period of seven months. Only 8 manhours were now used to unload the car. The continued savings and more efficient handling have no doubt enabled this brewer to continue to operate profitably. With the more widespread use of covered hopper cars in the transportation of bulk malt, the manhours per car has been further reduced, as has been spillage, accumulating additional savings.

For the small brewers, systems of 4-in. size having conveying rates of 450 bushels/hr, to systems of 6-in. size with conveying rates of 850 bushels/hr are sufficient for their needs. Large brewers find systems of 1000–2000 bushels/hr using 6- and 8-in. conveying lines suit their purposes satisfactorily. If unloading rates of greater than 2000 bushels/hr are required, it is much better to install two separate systems since a larger-sized system with its larger, or numerous hoses for connecting to the car becomes unwieldy; moreover, two cars can be unloaded simultaneously this way.

Today's breweries, like other manufacturing processes, call for maximum utilization of automatic controls, closely integrated flow lines, and high-speed conveying systems. Their size is continually increasing so that operating efficiencies must likewise increase, putting more emphasis on the handling of their raw materials. The pneumatic conveyor is practically a necessity to handle the raw materials, since it defies infestation of the grain through its self-cleaning property, maintains the kernel with minimum damage to its skin or jacket, and adds greatly to the physical appearance and cleanliness of the modern brewery.

The self-cleaning of the interior of the conveying system and the physical appearance and cleanliness of the brewery are evident by the wide acceptance pneumatic conveyors enjoy in the industry. By far the greatest objection to pneumatic conveyors is their ability to damage the malt during transit in the pipeline. This is only manifest when precautions to avoid damage are disregarded. By neglecting to recognize the necessary precautions and by not including them in the original design and installation, a less costly original installation will result, but this will reflect false economy. Breakage of the malt kernels and an undue large amount of skinning of the jackets will only result in difficulties in the process that follows, causing a reduction in production efficiency and greater variation in the final product.

The precautions to be taken when handling malt cover most aspects of a pneumatic conveying system. While they all have contributing factors to the whole operation, each should receive careful consideration, even to the point of impairing the physical appearance of the plant. They include the following:

1. Maintain a minimum pipeline velocity.
2. Use as dense a conveying stream as possible (low saturation, minimum cfm of air per pound of material conveyed).
3. Minimize number, and locate properly, conveying pipeline bends.
4. Avoid single blower combination vacuum-pressure systems.
5. Use proper feeders when metering the flow of malt out of bulk storage bins.

It is imperative that a minimum velocity be maintained within the pipeline system. Conveying malt at a high velocity is like putting it through a grinder, getting ground malt as the product. In the Saturation Tables for Vacuum Systems (Table 3.6 in Section 3.2), the velocity of 100 ft/sec is measured at the intake to the system. This includes the air leakage at the points of change from atmospheric to vacuum conditions. By using this 100 ft/sec figure in designing the air requirements for the system, we find that the velocity of the air in the system as it reaches the terminal end becomes 143 ft/sec when operating at a 9-in. Hg vacuum. It is difficult to believe that the malt attains this high a velocity since the thinner air at this vacuum has greater slip and lower impingement value. If a 12-in. Hg vacuum is used, the velocity of the air at the terminal end would be close to 170 ft/sec. As we can see from this, the higher the vacuum, the greater the velocity spread. The highest operating vacuum should probably be no greater than 10 in. Hg, which would give a terminal velocity of about 150 ft/sec.

It is interesting that European practice in such cases is to use stepped conveying pipelines that become slightly larger in size as the terminal end is approached. They have a distinct advantage since they have many more sizes of piping to choose from than we do here in the United States. Their amount of increase in pipe area is a very small precentage compared to the 4-, 5-, 6-, 7-, and 8-in. standard pipe sizes in the United States.

On pressure systems, the denser air allows a lower velocity to be used. The variation in the pressure system in velocity is not as severe as that in a vacuum system, but operating pressure should be no greater than 10 psig.

The number of pipeline bends in the system when conveying malt should be kept to an absolute minimum. Rectangular section bends should be used when possible so that wear to the pipe bend will not cause deep rivulets on the inside of the pipeline. The rectangular section bend should have renewable, abrasion-resistant wearplates having a thickness of no greater than $\frac{3}{16}$ in. In going from round to rectangular and vice versa, medium-length transitions should be used so that a gentle transition is made with minimum turbulence within the pipeline. If the location of the bend in the pipeline is such that the malt would follow the side plates of the rectangular section bend rather than the renewable back, then standard pipe bends should be used, preferably enamel lined (porcelainized).

As the pipeline approaches the terminal point of the system, its area should be progressively enlarged to reduce the terminal velocity. This enlargement, in the shape of the frustrum of a cone, should start far enough away from the terminal point so that the malt has just sufficient residual velocity to clear the end of the pipeline.

When pipe switches or diverter valves are used in the conveying pipeline, they should be able to allow a gradual change in direction rather than an abrupt change. The abrupt change will have the malt impinging on a surface that will produce severe ricocheting of the kernels, causing breakage. The flexible element type of pipe switch as described in Chapter 3 should be preferred.

If a combination vacuum–pressure system is used to convey the malt, always separate the vacuum and pressure legs into completely separate units. In other words, have a separate exhauster for the vacuum system, and a separate blower for the pressure system. If a single blower is used to activate both ends of the system, the variations in the amount of free air being handled by the vacuum leg of the system due to uneven loading in the system will be reflected in the pressure leg by much higher than necessary velocities. These higher velocities should be avoided.

In reclaiming malt from bulk storage bins, rotary feeders should not be used, since the kernels will be clipped between the rotor blade of the feeder and the body. Even though this clipping might be a low percentage of the total amount being conveyed, it is best to avoid the clipping completely. In such cases, open and closed type rotary valves should be used with an adjustable orifice in the sloping spout to the intake tee for metering the feed to the pipeline system. For grits, the rotary feeder is satisfactory. In pressure systems, while rotary feeders are necessary to seal the pressure in the pipeline against leakage to atmosphere, the feeder must be used, but in this case it should be nothing more than an airlock. Any buildup of material above the feeder must be avoided. The volumetric capacity of the feeder should be at least 1.5 times the conveying rate. A metered feed should be used prior to the rotary feeder to eliminate overloading the pneumatic pressure system as well as avoiding the clipping of the malt kernels in the feeder. In the throat of the feeder, a baffle plate should be used in order to divert the flow of malt away from the leading edges of the rotor where they meet the body.

The pneumatic conveyor is used in breweries to unload malt and grits, and, at times, rice from cars and to deliver to storage bins and reclaim from the storage bins and deliver to weigh hoppers, which in turn deliver the materials to the brewing process. In the small and medium-size brewery, both the unloading and reclaiming operations can be done through a single system (Figure 5.4). In this installation, the malt is unloaded from boxcars through the car unloading line, terminating in the filter above the storage bins. On discharging from the filter, the malt is passed through an automatic scale that will weigh the amount of malt in the car; it is then delivered to

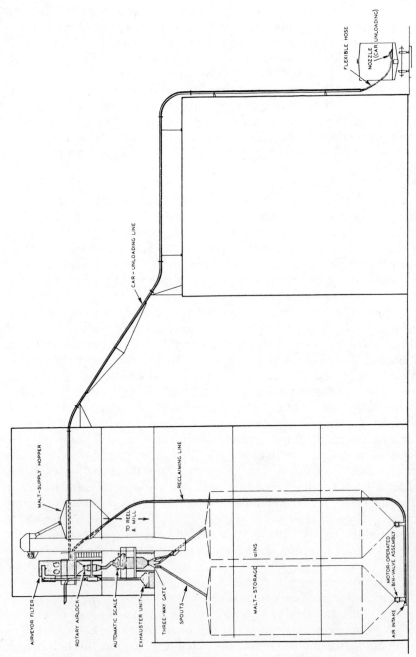

FIGURE 5.4. Combination unloading and reclaiming system handling malt. (Courtesy Fuller Company, Bethlehem, Pennsylvania.)

the storage bins through gravity spouting. When reclaiming from the malt storage bins, the malt is discharged from the bins through the motor-operated bin valves into the reclaiming line, again terminating in the filter above the storage bins. Through a counter on the automatic scale to register the number of dumps of the scale hopper, the correct amount of malt can be reclaimed for delivery to the malt supply hopper. The delivery from the scale discharge goes through a three-way gate and spouts diverting the malt to the bucket elevator, which in turn discharges into the supply hopper. From the supply hopper, the malt is delivered to the reel for cleaning and then on to the mill for grinding. Recently, the two-stage filter-receiver (Figure 3.10) has been substituted for the grain cleaner and the reel with a large degree of success.

In the larger brewery, a single system would be unable to maintain production, so separate systems are used for unloading and reclaiming (Figure 5.5). In this installation, the malt and grits are unloaded from cars from one of three unloading stations and delivered to any one of 14 storage bins. Through the use of a vacuum-type system from the cars to the unloading

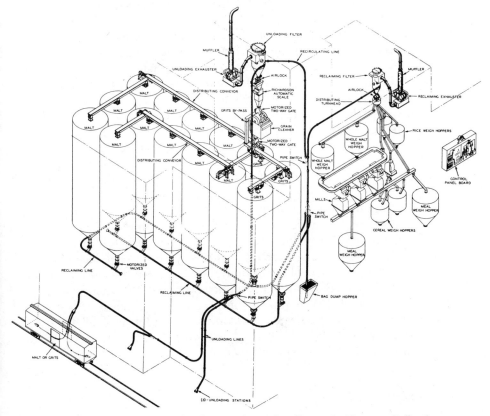

FIGURE 5.5. Separate unloading and reclaiming systems handling malt. (Courtesy Fuller Company, Bethlehem, Pennsylvania.)

filter, the malt and grits are conveyed horizontally and upward through a 180-ft ascent. From the discharge of the filter, again the automatic scale is used to record the weight of the incoming materials. In this instance, it was desired to clean the malt before delivery to the storage bins. The grain cleaner was installed below the receiving scale. From the cleaner, the malt is delivered to a series of mechanical screw conveyors for delivery to the proper bin. Since grits need not be cleaned, being a manufactured product, they are bypassed around the cleaner and delivered to their own set of screw conveyors for delivery to the proper bins. Note that this is a prime illustration of the use of two different types of conveyor to solve a complete problem, pneumatic conveyors to unload the cars and convey both horizontally and vertically to the top of the storage bins, and then screw conveyors to convey horizontally to the respective storage bins.

While the material is being unloaded, reclaiming from the storage bins to the brewhouse must also be done, so a separate reclaiming system is used. From the bottom of the storage bins, the motorized valves, on being opened, allow the malt to fall by gravity into the reclaiming line, where it is picked up by the conveying airstream and delivered to the reclaiming filter. The grits are fed into the conveying airstream by rotary feeders. From the reclaiming filter, the malt and grits are discharged through a swivel spout turnhead to weigh hoppers from which the malt, rice, and grits are delivered to their respective process equipment through mechanical conveyors.

In this industry, although two and sometimes three different materials are conveyed through the same pipeline, contamination between the materials is negligible. It is possible to stand some contamination of the malt by grits, but grits cannot stand contamination by whole malt. The whole malt kernels are seldom, if ever, found in the grits. To achieve this, however, when transferring from one material to the other the systems should operate for some 30 sec after conveying of material has ceased in order to clean out the system.

In the distilling industry, vacuum systems are used for unloading corn, rye, and malt from railroad cars and delivering to storage bins. Systems are installed having conveying rates as high as 2500 bushels/hr of the hard grains, with the malt being handled in the neighborhood of 3200 bushels/hr. The handling practices of the distilling industry follow those of the brewing industry. To the distilling industry, however, goes the honor of having the longest single-stage vacuum system. A Canadian distillery entrusted a pneumatic conveyor to convey their corn over a distance of 1400 ft from their storage silos to their distillery. The conveying rate has exceeded 20 tons/hr using a stepped conveying pipeline.

5.4 KAOLIN INDUSTRY

In the kaolin industry, pneumatic conveyors have both widespread and limited application, depending on the type of kaolin or clay being handled.

Their use is limited to that portion of the manufacturing process where the material is dry. This is during and after classification in the manufacture of air-floated clay; after drying in either a spray, drum, or rotary dryer for bead, flake, and lump clay; and after calcining and pulverization for calcined clay. Care must be exercised when spray-dried clay in bead form may be conveyed pneumatically since pneumatic conveying affects its particle shape and bulk flow. Some hard clays of South Carolina will have different bulk flow characteristics than the soft clays produced in Georgia. Moisture content will affect conveyability as well as the type of pneumatic conveyor to use.

Applications for pneumatic conveyors in the kaolin manufacturing plant that slurries its raw material from mine to mill include conveying the clay from the dryers and calciners to separators, screens, and storage, and from storage to bulk cars and bagging installations. The type of systems applicable to these operations include low-pressure, fluid solids pump, and the air-activated gravity conveyor.

Spray-dried clay is manufactured primarily to increase its bulk density and flowability and to suppress dust. The flowability increase, important to the manufacturer, is of extreme importance to the user. It reduces the user's problems when unloading the clay from cars since the spray-dried beads will flow much easier than the air-floated and waterwashed grades. The manufacturer, however, must maintain the bead form so that the user can expect minimum difficulty due to sluggish flow. When spray-dried clay is conveyed pneumatically, the beads are broken to the point where the clay approaches the pulverized state, which without fluidization is restrictive to bulk flow. While pneumatic conveyors can be used by the consumer, the manufacturer handling spray-dried clay should avoid them.

Air-floated clays, which are of low moisture content and are fluidizable and pulverulent, lend themselves to successful conveying by all types of pneumatic conveyors. Especially applicable are vacuum and low-pressure systems, and the fluid solids pump.

Waterwashed clays have higher moisture content than other clays. Only vacuum and low-pressure systems can be used, and with slightly higher velocities than those used in conveying air-floated clays. Normally, water-washed clays up to and including 3% moisture will convey at normal rates, but for each additional percentage point of moisture, there will be a 15% reduction in the conveying rate.

Kaolin, a clay for use in the ceramics industry, contains a rather high percentage of free silica. Whereas clays for the rubber and paper industry are considered nonabrasive, the clays for the ceramics industry are abrasive. The amount of free silica will determine the abrasiveness of the kaolin, as well as its conveyability with relation to the type of pneumatic conveyor to use. Normally, kaolin having up to 15% free silica can be handled in either vacuum or low-pressure systems using rotary feeders. Above 15%, gate locks should be used in place of the rotary feeder in such systems. This does not preclude the use of the fluid solids pump or the blow tank, no matter what the free silica content is, if the material is fluidizable and pulverulent. When the free silica is above 15%, these two types are ideally suited to the conveying of kaolin when that kaolin has the two physical characteristics mentioned. The blow tank can also be applied if the material is granular or lumpy.

5.5 PLASTICS INDUSTRY

It is difficult to state which had the greatest impact—the pneumatic conveyor on the plastics industry or the industry on the conveyor. The first pneumatic conveying systems were used in approximately 1930 for the conveying of phenolic resinoids, followed by wood flour, used as a filler for the molding of the resinoids into such items as telephone parts, radio components, and electrical insulators. The manufacturers of such resinoids were the first to apply pneumatic conveyors to their in-plant conveying operations between segments of the manufacturing process.

As other plastic materials were developed, again the manufacturers were quick to see the advantage of pneumatic conveyors and to use them most vigorously in their operations. These materials include vinyl resins, cellulose acetate, nylon, polyethylene, polystyrene, polypropylene, and other phenol, urea, and polyester compounds. When the molding and extrusion of these plastics increased to the point where these operations warranted bulk handling, pneumatic conveyors again were used. Between the manufacturer and the user, transportation methods had to be employed that protected the material from all hazards. This led to the development of bulk railroad cars and trucks that completely protected the lading from contamination from all sources.

In the conveying of plastics, special considerations are necessary. Contamination must be kept to an absolute minimum, not only from foreign matter, but also in the handling of different-colored materials such as polyethylene, which is molded and extruded in all the colors of the spectrum. One foreign, different-colored part in 80,000 may cause an off-color condition to the final product. Pneumatic conveyors, through their ability to be self-cleaning, minimize the possibility of such a hazard.

Temperature of the conveying air or other conveying medium must be carefully controlled, especially in pressure systems. In some instances, the heat of compression of the air or other medium in the blower will raise the conveying temperature of the air or medium to the point where it will affect the physical characteristic of the plastic being conveyed. Plastics are divided into two categories, thermoplastic and thermosetting.

Thermoplastic materials have the property of becoming plastic under the application of heat, are rigid at normal temperatures, and become plastic on each reapplication of heat. It is this type that requires controlled conveying medium temperatures so that the temperature remains at all times below the heat distortion temperature of the material being conveyed. Representative heat distortion temperatures, subject of course to variations, indicate the need for controlled conveying medium temperature:

Polyethylene	115–150°F
Polystyrene	165–200°F
Polyvinyl chloride	150–180°F
Nylon	165–170°F
Cellulose nitrate	110–150°F
Cellulose acetate	125–235°F
Ethyl cellulose	100–180°F

While these temperatures are the point where the rigidity of the material begins its transformation to the plastic state, the relatively short time the material is subjected to any higher temperature in the conveying system indicates that some higher conveying air temperatures may be used. Complete transformation is not instantaneous at the lower temperatures, yet surface change of the material may occur and affect its conveyability and its reaction within the conveying pipeline.

Thermosetting materials have the property of becoming permanently rigid under the application of heat, which originally is plastic or fusible, becoming infusible after the application of heat. When compounded with fillers for moulding, conveying temperatures are not as critical for this type of plastic. Heat distortion temperatures for these materials vary from 240–320°F, which includes the phenol, urea, and melamine formaldehydes.

Systems used in the conveying of plastics include vacuum, low-pressure, and the blow tank that uses low-pressure air requirements. Under no conditions should a fluid solids pump be used. The variable pitch of the screw will act on the material in the same manner as an extruder. In compressing the plastic, heat is generated whose temperature will be above that required for heat distortion, which means that instead of conveying dry material, the screw and its housing will become one homogenous mass of plastic. Higher pressure in the conveying system will also require more sophisticated air-cooling apparatus before the entry of the material into the pipeline, which in most cases becomes uneconomical.

Since conveying air temperature affects primarily plastic materials, we shall at this point indicate the method of temperature calculations for discharge temperature of a dry seal rotary positive blower.

1. Determine compression ratio:

$$cr = \frac{P_2 \text{ (discharge pressure absolute)}}{P_1 \text{ (intake pressure absolute)}}$$

2. Solve for Y factor as follows:

$$Y = (cr^{0.283} - 1)$$

This factor can also be found in tables in various handbooks, sometimes marked "values for X or for Y for normal air and perfect diatomic gases."

3. Solve for adiabatic hp as follows:

$$hp_{adiabatic} = 0.01542 \times \text{intake pressure absolute} \times \text{air volume} \times Y$$

4. Determine adiabatic compression efficiency as follows:

$$eff_{adiabatic} = \frac{hp_{adiabatic}}{0.95 \times bhp}$$

The 0.95 factor assumes the bearings and gear loss in the blower plus radiation is 5%.

5. Calculate temperature rise as follows:

$$T_2 = \frac{T_1 \text{ (ambient temperature)} \times Y}{eff_{adiabatic}} = °F$$

6. Calculate discharge temperature as follows:

$$T_3 = T_1 + T_2$$

As an example, let us design a simple system to convey polyethylene cubes at a rate of 60,000 lb/hr over a conveying distance of 400 ft.

1. From the saturation table (Table 3.10), we determine the saturation to be 1.1, the hp/ton to be 2.5, the velocity to be 70 ft/sec, and the pressure ratio to be 5.

2. Free air (scfm) = 1.1×1000 lb/min = 1100 cfm

3. Operating pressure = $\dfrac{2.5}{1.1} \times 5 = 11.4$ psig

4. acfm = 1100 cfm $\times \dfrac{14.7}{14.7 + 11.4} = 620$ cfm

5. pipe constant = $\dfrac{620 \text{ cfm}}{70 \text{ fps}} = 8.9$

Since the conveying pipe must be aluminum or stainless steel to avoid contamination, the pipe selected will be 5 in., schedule 5, which has a pipe constant of 9.4. This is close to 10% greater than the originally calculated constant, so we should recalculate the air requirements, based on this newer constant, as follows:

1. acfm = 9.4 × 70 fps = 658 cfm

2. psia = $\sqrt{\dfrac{1100 \text{ cfm} \times 14.7 \times (14.7 + 11.4)}{658 \text{ cfm}}}$ = 25.3 psia

   psig = 25.3 − 14.7 = 10.6 psig

3. scfm = 658 cfm × $\dfrac{14.7 + 10.6}{14.7}$ = 1132 cfm

4. Feeder leakage: assume 10% loss, 113 cfm
5. Blower air delivery: 1132 cfm + 113 = 1245 cfm
6. Blower hp = (1245 cfm + 20% for slip loss) × 10.6 × 0.005 = 79 bhp

We shall now calculate what the discharge temperature of the air from the blower will be when operating under an atmospheric pressure of 14.7 psia and an ambient temperature of 100°F (560°A).

1. cr = $\dfrac{(14.7 + 10.6)}{14.7}$ = 1.72

2. Y = $(1.72^{0.283} - 1.0)$ = 0.16588
3. $hp_{adiabatic}$ = 0.01542 × 14.7 × 1245 × 0.16588 = 46.8

4. $eff_{adiabatic}$ = $\dfrac{46.8}{0.95 \times 79}$ = 0.624

5. T_2 = $\dfrac{560° \times 0.16588}{0.624}$ = 149°F

6. T_3 = 100°F + 149 = 249°F

The discharge temperature of 249°F is considerably above the heat distortion temperature of 150°F. It is known that a pneumatic conveyor may tend to create traces of fines through abrasion of the polyethylene against the conveying pipeline wall, particularly if the conveying distance is long. These fines build up on the interior of the pipeline as film in the form of ribbons or streamers, or by names such as snake skinning and angel's hair. It is known that velocity control is a good solution in handling high-density polyethylene, but even the slowest allowable conveying velocity will have no effect when conveying low-density polyethylene. The newer linear low-density polyethylene is even more susceptible to the creation of fines, whatever the velocity. The most effective means of reducing fines—aside from buying fines-free resin—is to properly design the conveying system to have a minimum length, the least curvature possible (fewest pipe bends) and to treat the inner walls of the conveying pipeline for fines prevention. Treatment includes sandblasting, dimpling, and spiraling.

No doubt heat is also a contributing factor, but to what extent is a moot point. The author feels that as one of a number of contributing factors, it should be considered in the final design of the conveying system and its appurtenances. In designing a pressure-type system to convey plastics (not only polyethylene, but others) having a low heat distortion temperature, it appears advisable to use a maximum blower discharge temperature of some percentage of the difference between the heat distortion temperature and the molding temperature. Without the benefit of rigid laboratory tests to determine the degree of distortion during any increase in temperature, a figure of 20% seems to be a most likely figure. Representative maximum conveying medium temperatures for thermoplastic materials based on this hypothesis would be about as follows:

Polyethylene	160–200°F
Polystyrene	200–260°F
Polyvinyl chloride	180–225°F
Nylon	235–255°F
Cellulose acetate	170–285°F
Ethyl cellulose	160–245°F

By following this hypothesis, we can return to the polyethylene system previously designed where we had a discharge temperature from the blower of 249°F. While we may expect absorption in the equipment between the blower discharge and the material inlet, this will occur only until the equipment reaches the same temperature. There will also be some loss of heat through radiation, but for all practical purposes this should be ignored. We therefore find that the blower discharge temperature exceeds the maximum allowable conveying medium temperature by 49°F. This can be reduced to an allowable level by two methods: (1) the redesign of the system so that the blower discharge temperature is within the allowable conveying medium temperature; and (2) installation of an air cooler (heat exchanger) between the blower discharge and the material inlet.

To redesign the system, operating pressure must be reduced by enlarging the pipeline from 5 to 6 in., schedule 5, and calculate as follows:

1. acfm $= 13.5 \times 70$ fps $= 945$ cfm

2. psia $= \sqrt{\dfrac{1132 \text{ cfm} \times 14.7 \times (14.7 + 10.6)}{945 \text{ cfm}}} = 21.1$ psia

   psig $= 21.1$ psia $- 14.7 = 6.4$ psig

3. scfm $= 945$ cfm $\times \dfrac{14.7 + 6.4}{14.7} = 1356$ cfm

4. Feeder leakage: assume 10% loss, 136 cfm

5. Blower air delivery: 1356 cfm $+ 136 = 1492$ cfm

6. Blower hp = (1492 cfm + 20% for slip loss) × 6.4 × 0.005 = 57 bhp

To calculate for discharge temperature under similar conditions:

1. $cr = \dfrac{(14.7 + 6.4)}{14.7} = 1.435$

2. $Y = (1.435^{0.283} - 1.0) = 0.10761$

3. $hp_{adiabatic} = 0.01542 \times 14.7 \times 1492 \times 0.10781 = 36.5$

4. $eff_{adiabatic} = \dfrac{36.5}{0.95 \times 57} = 0.673$

5. $T_2 = \dfrac{560° \times 0.10761}{0.673} = 89°$

6. $T_3 = 100°F + 89 = 189°F$

This is within allowable limits, so it can be considered safe to use.

For air cooling, any standard heat exchanger, either tube or fin type, can be used. The housing, however, should be of heavy construction to withstand the discharge pulsations from the blower and should be equipped with a drain (Figure 5.6). Installation of the heat exchanger should be between the discharge silencer and the after-filter (Figure 5.7). This is a typical schematic arrangement of a blower installation. Air is drawn in through an intake screen and silencer to the blower inlet. The blower discharges into another silencer, which in turn discharges to the heat exchanger. To avoid vibration

FIGURE 5.6. Heat exchanger for blower discharge. (Courtesy Xchanger, Inc., Hopkins, Minnesota.)

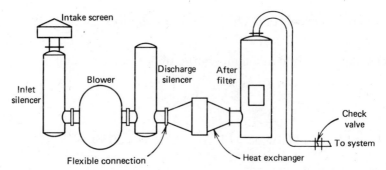

FIGURE 5.7.   Heat exchanger location.

transmission between the discharge silencer and the heat exchanger, a flexible connection should be used. The exchanger discharges into an after-filter from which clean air is delivered to the conveying system. The blower, silencers, and heat exchanger housing can all be made of ferrous metals; the after-filter can be made of aluminum or stainless steel, depending on requirements.

When we use an air cooler, a situation arises that is not readily recognized. In the design procedures, we use air at ambient conditions. The density or carrying capacity of this air depends on temperature, barometric pressure, and humidity (water vapor contained in the air). As the air is cooled, the water vapor is condensed and drained continually from the cooler. This loss of water vapor affects the density of the air and reduces its ability to impinge on and carry the material through the pipeline system. To replace this impingement value, air density in the pipeline must be increased. This can only be done by speeding up the blower sufficiently to replace the vapor loss through the cooler. A 15% increase in speed is a good rule-of-thumb figure to use. When doing so, however, recheck power requirements.

The previous discussion alluded to low-pressure, medium-velocity systems. What should the criteria be for medium-pressure, low-velocity (blow tank) systems? First, we must recognize that the material to be conveyed is in contact with the conveying air a much longer period of time than that in a low-pressure, medium-velocity system. Are the maximum conveying air temperatures then too high? Probably. It would seem reasonable then to use maximum conveying air temperatures equal to the heat distortion temperature.

Other points of concern in conveying polyethylene include avoidance of contamination, feeder clipping, and fines removal.

The avoidance of contamination is done by filtering all conveying air and by using nonferrous metals such as aluminum and stainless steel in construction. In vacuum systems only the intake air to the system needs to be filtered. In pressure systems, however, not only must the intake air to the blower be filtered, but the air on discharging from the blower must be filtered as well (Figure 5.7, less heat exchanger). This protects the conveying air from being polluted with rust or other particles being dislodged from the interior of the blower and discharge silencer; it is known as an after-filter. Without periodic inspections of this unit, any malfunction will be detected only in the final product, sometimes long after initial contamination has taken place. The air piping between the after-filter and the material inlet to the conveying pipeline system and the check valve should be aluminum to maintain cleanliness of the air after discharge from the after-filter. Where the polyethylene pellets come in contact with the conveying pipeline system, the construction of all components should be either stainless steel or aluminum.

When conveying preformulated resins, care must be used in the choice of either aluminum or stainless steel. Mineral fillers and extenders for resins include calcium carbonate, talc, kaolinite, aluminum trihydrate, feldspar, silica, glass spheres; agricultural by-product fillers include rice hulls, wood flour, and ground wood. Some of these fillers and extenders by themselves are abrasive. When they are mixed with both thermoplastics and thermoset compounds, do they alter these compounds in relation to abrasiveness and maximum conveying air temperatures? It would appear they do. But how much? Who knows. There have been reports of aluminum pipe showing an extreme amount of wear. When handling preformulated resins, stainless steel should definitely be considered.

For the elimination of feeder clipping in pressure systems, the side entry rotary feeder (Figures 3.16 and 3.17) is used. Through the side entry and controlled material flow, feeder pockets are only partially filled, completely avoiding clipping between the rotor and the body. When the feeder is used as an airlock below the cyclone receiver or filter-receiver of a vacuum system, the standard drop-through feeder with baffle plate and wiper blades will usually suffice.

Fines removal, depending largely on individual circumstances, should be made in the last component of the conveying system in which the material is conveyed. In large systems, fines separators of the aspirator type are the most prevalent and successful. For the processor, fines removal is done at the machine supply hopper. Small filter units with automatic air reversal for unit cleaning and two single-stage cyclone receivers are used. The receivers in series allows a complete separation of pellets and fines. The pellets are separated from the airstream in the first stage with the fines being carried in that airstream to the second stage. By discharging the second stage into a separate hopper, the fines are completely separated from the pellets.

In the manufacture of polyethylene pellets, vacuum, pressure, and combination systems are employed. They are used after pelletization has occurred and the material is dry. By applying pneumatic conveyors to the transfer between product bins and primary and secondary storage bins, to finished product storage for shipment, either by bags, containers, or railcars, the material handling and plant layout engineer is afforded a method of in-plant conveying that allows him to plan effectively the most economical plant arrangement. They also allow future expansion to be considered on a minimum cost basis. In considering what the future expansion will be and its effect on the conveying systems, it is wise to design the pneumatic systems on the basis of the foreseeable future expansion. The expansion many times involves only a longer conveying distance, which should be the basis for the present design. While this increases the original capital expenditure, it reduces considerably the cost of expansion, as well as maintaining production during the expansion period. If a system is to be extended, and its original design did not include the extension, the possibility exists that the system will be undersized for the extension in both power and pipeline size.

When conveying from storage bins, either vacuum or pressure systems can be used. The feeding mechanism for regulating the flow from the outlet of the storage bin into the conveying pipeline in vacuum systems is most advantageously accomplished through the use of orifice control, similar to that used for malt, or, if volume allows, through adjustable opening slide gates. On pressure systems, rotary feeders are used and can be attached to the storage bin since blowback air is dissipated quite easily through the voids between the pellets in the bin. Care must be exercised, however, to avoid clipping the pellets between the rotor and the body of the rotary feeder.

When delivering to storage bins, the pellets can be blown directly into the bin without the use of cyclones or other separating equipment. Vent stacks, of course, must be provided when using this procedure. Arrangement, size, and type of system to use will depend on the physical plant layout, number and size of storage bins, both primary and secondary, and production rates. In any event, do not be too conservative in determining conveying rates.

The users of polyethylene and polyolefin pelleted resins find bulk handling a most effective cost-reducing operation. The use of the covered-hopper railroad car designed to protect its lading from all contaminants, including dirt, cinders, rain, hail, sleet, and snow, enables them to save $0.04/lb by receiving their resin by these cars in bulk rather than in bags. On a carload (100,000 lb.), this means a direct saving of $4000.00. Add to this a saving of $0.015/lb in labor costs when using pneumatic conveyors to unload the car, store the material, and deliver it to the extruders, sheeters, or molders and the total saving becomes $5500.00/carload. This saving is afforded not only the large user but the small user as well. Most producers and suppliers maintain their own bulk cars either through direct ownership or by a lease arrangement. Since these cars are void of standard railroad demurrage charges, the suppliers set the rules, which normally allow the user to keep the car in his plant at no charge for a period of 14 days. This means that the user can use the car as his main bulk storage bin, minimizing the amount of bulk storage he himself must maintain, producing further savings in his costs. Variations will occur in arriving at the minimum consumption that is required to consider bulk handling, but a good figure is 15,000 lb/week.

When the cars are used for storage, and only one type of polyethylene resin is used to serve only one processing machine, a simple vacuum system (Figure 5.8) is all that is necessary. Conveying rate or unloading capacity for such installations will vary from 1000 to 3000 lb/hr. Design of such systems with relatively low conveying rates requires that saturation figures be increased perceptibly since they will use small conveying pipeline sizes, which result in higher frictional resistances. For such systems, saturation should be increased at least by 100%. When the number of processing machines is increased to two or more, a combination vacuum-pressure system (Figure 5.9) should be used. Conveying rates for such systems will vary from 2000 to 5000 lb/hr. Again, saturation figures should be increased, but by a minimum of 75%. On many such systems, the conveying distance of the vacuum leg is extremely short. Under this condition, the only practical guide is experience. Illustrative of this is the saturations achieved over a 30-ft vacuum conveying distance. When using 3-in. hose, a conveying rate of 10 tons/hr can be achieved, with a saturation of 0.65 cu ft of air per pound of material conveyed; when using 4-in. hose, a conveying rate of 20 tons/hr can be achieved, with a saturation of 0.56 cu ft of air per pound of material conveyed.

When a variety of polyethylene resins, either high density, low density, or colored, is used, it is best to use a combination vacuum-pressure system (Figure 5.10). Unloading rates, where required, can approach 50,000 lb/hr. Since this arrangement is similar to the previous one, again a vacuum system is used to unload the cars and deliver to a transfer point, from which a pressure system is used to deliver to a multiplicity of discharge points. In the pressure system, two methods are employed, single pipeline and multiple pipeline. When the variance in grade of the polyethylene pellets being handled is the only criteria, the single pipeline system with flow-diverter valves

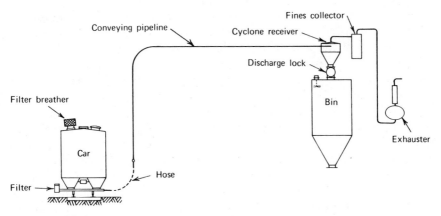

FIGURE 5.8. Simple vacuum system unloading polyethylene cubes.

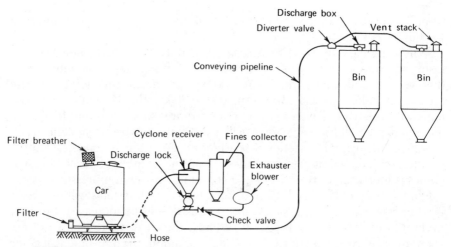

FIGURE 5.9. Combination vacuum-pressure systems unloading polyethylene cubes.

is possibly the best arrangement. When other variances, such as color differences, are encountered, the multiple pipeline system is best to avoid contamination between colors. Another determination that warrants consideration is the amount of automation desired or required. The flexible hose switch and multiple pipelines is excellent for manual operation. Automating a multiway diverter valve or pipe switch can produce quite a complex piece of machinery, and an expensive one. In the interest of maintenance, however, the flexible hose switch is very reliable and easily maintained. Since it is usually placed in the most accessible location, if maintenance is required, it is a simple operation. In the case of the single pipeline with flow-diverter valves, although the valves are easily automated and afford simple interlocking through limit switches for system control, their location above the bins, where accessibility at times is limited, is subject to more extensive maintenance procedures.

In many cases, a single blower is used to activate both the vacuum and pressure system. Many systems are operated successfully in this manner, but if we accept the hypothesis outlined earlier regarding maximum heat distortion temperatures and their effect on the material being conveyed, air discharge temperature to the pressure system should be checked. Separate blowers to operate the vacuum and pressure systems should be considered if the air discharge temperature is alarmingly high.

To deliver the polyethylene resins from the storage bins to the processing machines, pressure systems are used (Figure 5.11). As can be seen from this arrangement, we deivate from the criteria that pressure systems are used in conveying from a single pickup point or intake point to a multiplicity of discharge points. Often we can and do have more than one pickup or intake point. We can do this because (1) we are conveying a relatively dust-free material that can be fed into the conveying pipeline through a rotary feeder

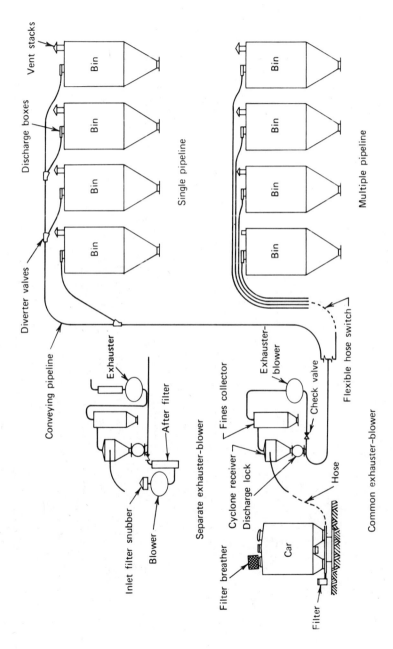

FIGURE 5.10. Combination vacuum-pressure systems unloading a variety of pellets.

173

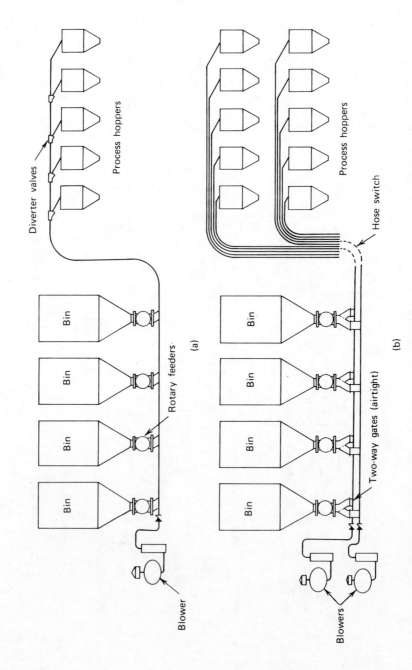

FIGURE 5.11. Low-pressure system, storage bins to processing hoppers. (*a*) Single pipeline, single system. (*b*) Multiple pipeline, dual system.

attached directly to the bin outlet and gate without the blowback of air through the feeder impeding the flow out of the bin and into the feeder pockets, (2) the material requires a velocity that causes sufficient air to be delivered to the conveying pipeline for conveying, and (3) minor air losses can be absorbed through the expansion of the conveying air as its pressure decreases.

Systems such as this, whether in the producer's or user's plant, are easily automated and controlled from electric control panels. High- and low-level indicators in the processing machine hoppers indicate the need for material at these points, and through relays and automatic starters and switching equipment, systems can automatically satisfy the requirements.

In the user's plant, where a single grade of material is used and a multiplicity of process machines are to be served, a valveless conveying system might be employed. This system (Figure 5.12) utilizes a scalper tank above each machine, inside which the conveying pipeline is broken so that the material being conveyed will fill the first tank and then each succeeding tank. As material is withdrawn from the tanks, it is promptly replaced by the scalping action within the tanks. This type of system can be either continuous or intermittent governed by bin level indicators in the scalper tanks.

Polyvinyl chloride was first conveyed pneumatically by the manufacturers in their in-plant conveying from process to storage and to shipping facilities. This material is one whose physical characteristics can vary considerably, so care should be exercised in the choice of the components used in the whole system. Its fineness will vary from all passing through a 325 mesh screen, to spheroids approximating 100–200 mesh size, to coarse material where it is all below 38 mesh size, and to pellets of ¼-in. size. With this variation, it is not surprising that its bulk density will vary from 17 to 50 lb/cu ft. While the material is known not to support combustion, its travel

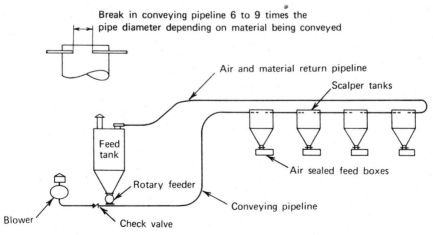

FIGURE 5.12. Valveless scalper system for plastic pellets.

through the conveying pipeline causes considerable static electricity, so grounding of all equipment to dissipate this static is necessary. This will prevent sparking, which would cause explosions when such atmospheres exist outside the conveying pipeline and it will protect the operator against static shocks.

This static phenomenon was discovered very much by accident by one of the early producers of polyvinyl chloride. This producer installed quite an elaborate pneumatic conveying system, which had a successful startup and maintained satisfactory operation. Operation at first was limited to daylight hours. Production was on the increase so night operation became necessary. During the second night of operation, the plant engineer received a frantic telephone call from the system operator at the plant. "Sir," he said, "you know that PVC system, something is happening. It operates perfectly, but up in the ceiling where the pipes goes across, there is an aurora about 18 in. in diameter, and about the prettiest sight you could see. I thought you ought to know." Well, the plant engineer surmised what was happening and immediately told the operator to shut the system down. The engineer then returned to the plant and put the system back into operation. After several hours the aurora reappeared, much like the aurora borealis. This convinced him that the cause was static electricity and the system was shut down for the night. Next day, the entire system was grounded, including jumper wires across the conveying pipeline joints. As in all good stories, the system lived happily ever after. Fortunately, there was no loss of any kind, except a good night's sleep.

Vacuum systems as well as pressure systems of the blow tank type are used in conveying polyvinyl chloride. Some pressure systems using rotary feeders have been successful with much care emphasized in the design and constructiofn of the rotary feeder. Polyvinyl chloride, being very sensitive to heat distortion, requires that the rotary feeder, whether used in vacuum or pressure service, be constructed with such rigidity that the clearance between the rotor and the body of 0.002 in. be maintained and requires that it be perfect as possible. This minimizes the possibility of the material rolling between the rotor blade and the body, with the ensuing friction generating sufficient heat to cause particle distortion. The feeder rotor blades should be back-beveled at their perimeter and ends and surfaced with a material having a slightly lower hardness than the body to prevent galling. Feeders should also be direct connected to their driving mechanism through a flexible coupling, not through sprockets and a roller chain. By direct connecting, chain pull, the cause of most of the distortion in a rotary feeder, is eliminated.

In vacuum systems, the filter-receiver should have an adequate amount of filtering media. Polyvinyl chloride is a material that does not build up a filter cake on the media, so the amount or area of the media is important to maintain a low velocity through it. Low velocity will avoid bleeding through the media, especially on the spheroid types.

Phenolic resin, a thermosetting material, is conveyed pneumatically by both the user and manufacturer with both vacuum and low-pressure systems being employed. When conveying phenol formaldehyde exclusive of any filler, the conveying air temperature should not exceed 100°F, since the compound has a heat distortion temperature of 100–260°F. When it contains fillers such as wood flour, mica, asbestos, and fabric, its heat distortion temperature varies from 240 to 350°F. Other thermosetting materials such as phenol-furfural, urea-formaldehyde, melamine-formaldehyde, and polyesters have heat distortion temperatures of 240–320°F.

In the manufacture of phenolic resin, when it is formed into the dry state, it is in a granular and lump form. For molding use, it must be in powder form, so it is passed through a grinder to attain that form. For conveying from the grinder to process or storage bins, vacuum systems are used universally. All of the conveying air is drawn through the grinder to air-sweep the machine. Under this condition, the amount of air required to air-sweep the grinder is much more than that required to convey the amount that has passed through the grinder. Hence the pneumatic conveying system is designed not by the minimum requirement of the cubic feet of air per pound of material conveying per minute (saturation) but by the amount of air that is necessary to air sweep the grinder. If this is not done, an overheated grinder will result.

The filter-receiver in a vacuum system handling thermosetting materials should have a very low air-to-cloth ratio. The ratio for woven and felted filter media should be a minimum of 2:1 and 3:1, respectively. The filter-receiver should be equipped with explosion vents and the media with static wires to dissipate static electricity. All parts of the system should be properly grounded. Rotary feeders in both vacuum and pressure systems should have close clearances between the rotor and the body and should be subject to minimum distortion. Materials of construction, ferrous or nonferrous, will depend on the material being conveyed and its end use.

Wood flour, one of the most used fillers for phenolic molding compounds is conveyed mostly by vacuum-type systems. Low-pressure-type systems are used only where the number of discharge points would require considerable mechanical or other conveyors, including dust suppressors, to reach a good, operational arrangement, with minimum hazards. Wood flour is an explosive material. It is therefore most wise to design bulk handling systems adhering to specifications developed by organizations such as the National Fire Protection Association.

The transportation of bulk wood flour is limited to van trucks and railroad boxcars. Since it is a material whose flow charcteristics are 100% negative, covered hopper and specialty railroad cars and trucks simply cannot be used. Wood flour will not flow out of such cone-bottom facilities. Large bulk storage facilities must be flat bottom with a rotating rake to facilitate discharge. Smaller use bins can be rectangular in shape with opposite ends vertical, with the sides having a slight slope to a twin screw discharge, plus vibration. At the discharge from any type of bulk storage bin, a rotary feeder should be used as an explosion choke to prevent an explosion exterior of the bin to penetrate into the bin through the discharge mechanism. All equipment must have properly sized explosion vents and must be adequately grounded to dissipate static electricity. Spout magnets also must be used to prevent tramp iron from causing sparks in the explosive atmosphere.

A wood flour bulk handling installation is illustrated in Figure 5.13. This installation was constructed for a large molder of phenolic resins who found it more advantageous to blend his own compounds than to purchase pre-blended materials. The usage was more than sufficient to justify bulk handling and storage rather than bagged material. The wood flour was received in boxcars, some of which traveled 3000 miles from the wood flour processing plant to his railroad siding. After traveling these many miles, the wood flour was fairly compacted within the car and manual assistance to the vacuum type system was required to get the material out of the car. With the material being received in boxcars, the vacuum system was ideally suited to suck the material out of the car. After unloading the material from in back of the bulkhead, which spanned the door of the car, the operator placed the nozzle on the floor of the car. In this manner, he was able to break down the pile, undercut it, and take advantage of pile crumbling and submersion of the nozzle in the pile to reduce the amount of manual labor involved. The hose and nozzle were constructed of steel and connected to a standard steel, Schedule No. 40 conveying pipeline. While the conveying pipeline was grounded to dissipate static electricity, it was found necessary to dissipate more directly the static electricity generated in the flexible metal hose and steel nozzle. This was done by means of grounding the flexible metal hose to the railroad track. The railroad car, having wood sheathing and flooring, was insulated from the railroad track, so the wood flour lading inside the car was also insulated. The grounding of the flexible metal hose dissipated the static electricity and avoided any possible explosion within the railroad car and exposure of the operator to static shocks.

The wood flour, as it was unloaded from the boxcar, was conveyed through the standard steel pipeline whose joints were joined with static jumper wires, and terminated in a filter-receiver above the storage bins. This filter-receiver was equipped with explosion vents. The cloth filter tubes within the filter-receiver were equipped with static dissipating wires woven into the seams of the tubes. The static wires, in turn, were attached to the necks of the crown sheet so that the static collected in the tube was dissipated through the crown sheet and the subsequent grounding of the filter-receiver to the entire bulk handling and storage installation.

As the next step in Figure 5.13, the wood flour was separated from the conveying airstream in the bottom cone of the filter-receiver and discharged through its bottom outlet. In order to facilitate the flow out of the filter-receiver into the rotary feeder pockets, the feeder was constructed with straight necks, both inlet and discharge. This avoided any restriction to the

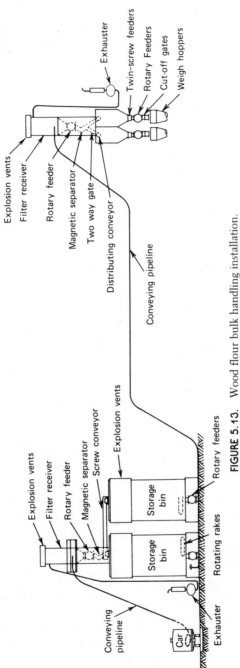

FIGURE 5.13. Wood flour bulk handling installation.

flow of material. From the discharge of the feeder, an offset spout with permanent magnet type magnetic separators was installed to take out tramp iron such as nuts, bolts, nails, and wire from the incoming stream of material. Thus this material was prevented from entering the storage bin where its possible sparks could cause an explosion. For delivery to the storage bins, a horizontal screw conveyor was used with its end bearings exterior to the screw conveyor trough, and the interior bearing hangers equipped with lignum vitae bushings to minimize friction at these points, which could cause heat for possible combustion.

The storage bins were flat bottom and used a rotating rake for discharging the material. The bins were constructed of welded steel plate. The top two courses of plate were constructed as explosion vents. Such large explosion vents were necessary in order to adhere to the rigid specifications employed. In some cases, the roof of such a bin is lightly fastened to the sides so that the entire roof becomes an explosion vent. If an explosion occurs, the roof would be blown off rather than the sides blown out. In this case, with equipment installed above the roof, it was deemed advisable to put the explosion vents in the side of the bins.

The installation also conveyed the wood flour from the storage bins to the molding plant over a conveying distance of approximately 400 ft. While it is known that vacuum type pneumatic conveying systems offer little assistance to the propagation of flame and explosion in the conveying pipeline, it was decided that as an added safety feature the storage facility would be isolated from the molding plant. This was done by installing a rotary feeder between the discharge outlets of the rotating rake mechanisms and the intakes to the vacuum reclaiming system.

At the terminal end of the reclaiming system, another filter-receiver is used having the same protective devices as that above the storage bins. The entry spouts into the two use bins were equipped with spout magnets as a further precaution against tramp iron. Material distribution within the use bins was by open helicoid screw conveyors with all bearings exterior of the bins. High-level indicators of Class II, Group G construction was installed at the tail end of the screw conveyors to shut off the rotating rake discharging mechanism in the storage bin when material reached its rotating paddles. Through time delay relays, the system was allowed to operate for another 30 sec to clean itself out.

For discharging the wood flour from the use bins, twin screw feeders were used for delivery to weigh scales and processing equipment. At this point, it is highly recommended that rotary feeders be used between the outlet of the screw feeders and the inlet to the weigh scales. These would further isolate the volume of wood flour in the use bins from the operating and molding area.

In installations such as this, static jumper wires are placed across all pipe joints in the conveying pipeline system, as well as the grounding of all equipment. All electrical equipment must be explosion-proof. When constructing systems such as these for the conveying and storage of combustible materials, it is wise to check with the insurance company and the underwriters to be certain that the installation and its safeguards meet with their every requirement.

5.6 RUBBER INDUSTRY

The application of pneumatic conveyors in the rubber industry is widespread. The in-plant conveying of rubber pellets from pelletizers and cooling conveyors to curing stacks and compounding stations is most adequately done by vacuum systems. These operations are mostly batch-type operations that lend themselves to the complete conveying of the batch without residue.

As the rubber pellets are discharged from the pelletizer and passed through the cooling conveyors, they are covered with soapstone to avoid their sticking together. It is this soapstone that requires the terminal end of the vacuum system to be equipped with a filter-receiver. The amount of filter media in this filter-receiver should be generously proportional to the conveying air, since the soapstone is a fine material that can easily bleed through the filter media if the velocity through the media is excessive. In some cases, it is desirable to separate the excess soapstone from the pellets at this point. To do this, the two-stage filter-receiver is used where the rubber pellets with the normal amount of soapstone adhering to them are discharged through the primary outlet, and the soapstone dust is discharged separately through the inner cyclone and secondary outlet to be reused in the process.

From the cooling conveyors, rubber pellets are discharged into a movable plate feeder to maintain separation of the pellets and flowability. The pickup to the vacuum system is normally just the conveying pipeline itself, no nozzle jacket. The end of the pipe is placed approximately 2–3 in. above the top level of the pellets on the feeder.

With the operations being of the batch type, the batches at the terminal end of the system are retained in two types of receptacles prior to feeding into the banbury mixers. Where single banbury mixers are serviced, the holding of the batch of rubber pellets is done either in the bottom cone of the filter-receiver or in an inverted truncated spout below. Where more than one banbury mixer is serviced by one conveying system, the pellets are discharged either into weigh larries or two or more inverted truncated spouts. In the case of the spouts, they are rectangular or square in cross section with wider dimensions at the bottom than at the top. When the outlet gate at the bottom is opened, this gives a free fall without restrictions.

Below the filter-receiver, a simple open-and-shut type of slide gate is used instead of a rotary feeder to shut off discharge so that vacuum and the airstream can be distributed to the conveying pipeline system for conveying.

One thing must be certain: the volumetric capacity of the bottom cone of the filter-receiver must be sufficient to hold the largest batch that is to be conveyed at any one time. With the batches being transferred at a rapid rate, the exhauster that activates the system is allowed to run continuously during the operating period. Constant starting and stopping the exhauster driving electric motor will put excessive strain on not only the windings of the motor but the starter as well.

With the exhauster running continuously, a vacuum breaker must be installed in the airline between the filter-receiver outlet and the intake to the exhauster so that the air to the exhauster is drawn in through this vacuum breaker. This will reduce the vacuum load in the filter-receiver and avoid the draw through of air in the pipeline system. After the vacuum is reduced, the open and shut gate will operate easily and not have to work against an operating vacuum.

In setting up a compounding installation, care must be exercised in the timing of each operation, especially when it is automated. When weighing operations are included, and are tied into pneumatic conveying systems, time must be allowed for weighing, stabilization of the scale-weighing mechanism, and discharge. Time should also be allowed in the conveying cycle to take care of the complete cleaning out of the system after each conveying operation. To determine the conveying rate of the system, time should be allowed for the initial start of the conveying operation, the length of time allotted to convey the complete batch, and the time required between cessation of conveying and the cleaning out of the system.

In designing vacuum systems for rubber pellets, the saturation figures in the Saturation Table (Table 3.6) are for in-plant systems. The minimum size conveying pipeline for conveying rubber pellets should be 4 in. No correction factors should be used for systems having a pipeline size greater than 6 in. up to and including 10 in.

In the manufacture of rubber tires for automobiles and other automotive equipment, as well as other rubber products, carbon black is used as a filler. The carbon black, when handled in bulk, is received in specially constructed railroad covered hopper cars. The carbon black is pelletized by the producers to facilitate handling and to reduce dust in transferring from one place to another. The pellets, however, are quite fragile. Early attempts were made to convey this carbon black pneumatically from the outlet of the railroad covered hopper cars to storage bins. While the pneumatic conveying system operated satisfactorily, the degradation to the pelletized carbon was another matter. In conveying through a vacuum type system, the pellets either exploded when subjected to the negative pressure of the system or were broken by attrition within the conveying pipeline. When the carbon reached the storage bin, it was not pelletized but pulverized. In this state, it was found to be very difficult to get out of the storage bins, and, further, the dustiness in this state caused quite a few problems in conveying the carbon from storage bins to weighing and compounding stations. This carbon black should be handled as gently as possible in order to retain the pelletized state. Mechanical conveyors such as horizontal screw conveyors and vertical screw lifts should be used.

With the use of rubber expanding to many household items, and its replacement of leather for soles and heels of footwear, other filler materials have come into widespread use. One of these materials is rubber-grade air-floated clay. When the amount of clay being handled and used warrants bulk receipt and handling, vacuum and low-pressure systems and fluid solids pumps are used.

This type of clay is transported in covered hopper railroad cars. Although it compacts during shipment, it can be unloaded by either a vacuum type system or fluid solids pump quite easily through the application of vibration to the car to maintain flow of the clay out of the car and into the system. In using a vacuum system, maintaining a flow of clay out of the car is not difficult, but it does affect conveying rates. If a fluid solids pump is used to convey from the car to storage, flow must be maintained at all times. Level indication in the pump hopper to control operation of the pump screw only when material is available to it will alleviate many operating problems.

For storage of this air-floated clay, welded steel bins with smooth interior welds should be used. In order to maintain a minimum amount of friction between the clay and the walls and bottom cone of the bin, the bin interior should be treated with any one of the special coatings that are available to make a smooth surface. Two coats of spar varnish after wire brushing will do a good job.

The bulk density of this clay is seriously affected by aeration. In designing equipment for delivery from the storage or use bins to weigh hoppers, extreme care must be exercised. A method that has been used successfully is a two-speed screw conveyor plus a fast-acting cutoff gate directly above and as close as possible to the weigh hopper. In metering the amount of clay into the banbury mixers, a high degree of accuracy is desired. Through controls on the scale dial, when feeding the weigh hopper commences, the screw conveyor operates at its highest speed. As the weight of material in the weigh hopper increases and approaches the desired weight, the screw conveyor is then reduced to half speed for a dribble feed to the weigh hopper. When the material in the weigh hopper reaches the correct weight, the control on the scale closes the quick-acting gate directly above the weigh hopper and shuts down the screw conveyor. Some minor adjustment to the scale controls will be necessary to take into account the amount of clay in suspension at time of cutoff between the quick-acting gate and the level of clay in the hopper. The volumetric capacity of the weigh hopper should be based on the clay being in an aerated condition. While this rubber-grade clay normally has a bulk density of 32–35 lb/cu ft, the weigh hopper should be designed for a bulk density of 20 lb/cu ft.

5.7 PULP AND PAPER INDUSTRY

Pneumatic conveyors have an extremely wide application in pulp and paper mills for the handling of dry bulk chemicals, wood chips, bark, sawdust, hog fuel, and dried pulp. In addition, pneumatic conveyors have application for the conveying of wood chips and sawdust in sawmills and veneer and plywood plants, which supply a large portion of wood requirements to pulp mills.

The modern pulp mill receives pulpwood in the form of chips, sawdust, and logs. The chips and sawdust are produced from lumber and plywood residuals, and received by the mill in railcars, trucks, or barges. The logs are barked prior to chipping. After chipping, the chips are collected for delivery to storage.

After the chips and sawdust are unloaded by mechanical means and the chips are collected from the chippers, low-pressure pneumatic conveying systems take over the handling problem. These can even do some chip processing. One such installation conveys the chips and sawdust to outdoor storage; reclaims from storage through multiple feeder inlets for accurately controlling blends of wood species, and sawdust and chip combinations for delivery to screens; and conveys screened chips and sawdust to the digesters.

Many pulp mills are being, have been, and will be constructed utilizing as their only source of wood fiber chips purchased from sawmills and veneer and plywood plants. The sawmills as they saw the logs into structural timber produce residuals such as slabs and edgings. Before the 1950's, these slabs and edgings were discarded as firewood. With pulp mills accepting this wider source of supply for their wood requirements, the sawmills found a market for these slabs and edgings, provided they could debark and chip them in such quantity that the pulp mill would be receptive to this wood supply.

After the sawmill passed the slabs and edgings through a chipper, a method of getting chips from the chipper into conveyances such as trucks and railcars for transportation to the pulp mill was required. Pneumatic conveyors, requiring only a pipeline for the transportation of chips to the conveyances, found a ready application. Low-pressure-type conveying systems were developed along with special discharge nozzles for loading and compaction of chips in vans and railcars. Such systems are simple and contain a positive-pressure blower unit, a rotary feeder of fabricated construction rather than cast, conveying pipeline, and the car-loading nozzle.

The veneer and plywood plants peel the logs into thin sheets, $\frac{1}{2}$ in. thick and under, to the point where a core of about 5–6 in. in diameter remains in the peeler lathe. Complete wood utilization is accomplished by sawing this core into 2×4 structural lumber and then chipping the edgings from this sawing operation into chips.

After transportation by either trucks or railcars to the woodyard of the pulp mill, the chips had to be unloaded and a new type of storage developed, outside chip storage. The pulp mill now had to go to a material-handling technique different from that used when handling roundwood. They now had a material that could be handled and conveyed very easily and economically in bulk through pneumatic conveyors.

The handling of wood chips by pneumatic conveyors is not a recent development. Fifty-five years ago, a low-pressure-type system handling wood chips was installed in a pulp mill in Ogdensburg, New York. This system conveyed wood chips from a rotary screen in the woodroom to the chip loft above the digesters over a distance of 450 ft. It conveyed the chips at the rate of 15 tons/hr through a 9-in.-diameter sheet steel conveying duct, including seven 9- × 7-in. rectangular section bends, aggregating a total curvature in the pipeline system of 450°. A positive-pressure blower was employed, driven by a 50-hp motor. The maximum operating pressure of the system was 3 psig. By today's standard, this would be called an extremely small installation.

In the woodroom of today, the wood chips after being formed in the chipper are also passed over screens to remove cards and slivers as well as sawdust. With high production rates, and with the screens being placed close to the chipper, the chips are blown to a cyclone receiver above the screens. This operation is done by using air that is generated by the chipper and the attachment of fan blades to the chipper head to give added air for velocity and pressure increase. The maximum conveying distance for this operation is about 200 ft.

From the screens, the chips are discharged to a collecting belt conveyor, which in turn delivers the chips to a single discharge point for feeding to a low-pressure system. With outdoor chip storage, it was only natural that pneumatic conveyors received prime consideration for the handling of chips for such installations. The pneumatic conveyors of the low-pressure type can convey wood chips over distances up to 5000 ft and can distribute to any number of chip piles, building up the piles so that it appears that mountains of chips abound at the pulp mills.

It is not the purpose here to delve into the relative merits of outside chip storage. Much research is under way regarding the changes in composition that can occur while chips are stored outside, improving control of chip size, and reducing the loss of yield in cooking. The purpose here is only to note the appropriateness of pneumatic conveyors to the conveying of wood chips for such installations.

In some woodrooms, the chips are conveyed directly to storage or digester supply bins without screening. When using pneumatic conveyors to convey from the discharge of chippers, systems can very easily be undersized. Instances of this type have been reported, caused primarily by the lack of appreciation of extreme variations and surges in chipping rates, and wood densities and moisture. Frequently this is not the fault of the pneumatic conveying engineer alone. It is the fault of the woodroom designer in setting conveying rates, and choice of allied equipment.

A good illustration of this is a pulp mill installation that used a low-pressure-type pneumatic conveying system to convey the chips from the discharge of the chipper with a surge hopper approximately 8 ft high between the bottom outlet of the chipper and the inlet to the rotary feeder of the conveying system. The information given the conveyor engineer was that the chipper would produce 62½ tons/hr with surges up to 75 or 80 tons/hr. From this information, three quotations were secured. Each specified a different pipeline size and horsepower requirement. The pipeline sizes were 8, 10, and 12 in., and the horsepower requirement was 100, 100, and 125, respectively. Their prices were in the same relative position as their pipeline size.

The conveying engineer of this installation, who recommended the 12 in. pipeline, questioned the conveying rate given. The chipper that was purchased could chip at a greater rate than the specified surge rates of 75 or 80 tons/hr. The chipper was to make ⅝-in. chips and to chip logs up to and including 14 in. in diameter. If, for a certain period, all logs being chipped were 14 in., of 4-ft sticks (20 sticks per cord), the normal continuous feed to the chipper would be 50 cords/hr. Figuring 5000 lb of chips per cord, there would be 125 tons of chips per hour being delivered to the conveying system, 50% more than originally specified.

The small-sized systems would have had extreme difficulty in conveying away the full potential output of the chipper. If this were not done, and chips backed up through the surge hopper into the chipper itself, there would have been one unholy mess. Fortunately, the 12-in. system was installed on the honest appraisal and knowledge of the conveyor engineer who recognized the variations involved. As far as is known, no backup of chips into the chipper has ever occurred.

During the past 10 years, chip breakage and fractionation (propagation of wood fines) in pneumatic conveyors has become a matter of concern to quite a few pulp mills. We must recognize, however, that in the present methods of outside chip storage, the chips are subjected not only to at least one pass through a pneumatic conveyor, but also to external forces such as bulldozers used for spreading out the piles for storage and collecting the chips to reclaiming points. To minimize breakage as well as fractionation in the penumatic conveyor, velocity within the conveying pipeline should be closely controlled, pipeline joints should be checked periodically for misalignment and abutment, and most important, pipeline rectangular section bends should be inspected for internal wear. This internal wear can be the major contributor to the problem.

Unfortunately, the woodyard in most mills is the last-priority area for maintenance. With this in mind, we should ask, "When was the last timne the pneumatic conveyor pipeline was inspected?" Was it last week, last month, or six months ago, or was it when holes appeared in the wearbacks? To keep operating with minimum interruption probably the welder went out with some patches and welded them on the exterior to seal the holes.

If this last procedure was done, a roughened interior became worse instead of better, furthering breakage and fractionation.

How do we minimize the problem. First, a maintenance schedule should be set up to inspect the wearbacks and transitions of the pneumatic conveyor. After this, how do we determine internal wear without taking the system apart? Simply by using an ultrasonic thickness gauge.

Before we can use the gauge, we must find where the points of wear might be. After the conveyor has been operating for a couple of hours and conveying chips during that time, test for hot spots. In any bend, material will impinge at certain areas with the friction between the chips and the bend causing the area to increase in temperature. Starting at the transition at the point of curvature, run your hand along the length and width of the bend and its transitions. It might also be a good idea to continue for some distance along the round pipe after the point of tangency of the bend and the transition. At each warm or hot spot, mark the area with a yellow lumber crayon. When conveying is completed and no material is being conveyed, take the thickness gauge and go over the entire areas marked with the crayon. If you find that thicknesses vary considerably, there is a good possibility that the interior of the bend is worn and may contain many rough spots that contribute to breakage and fractionation. Replacement of the wearback will be necessary so that a smooth inner surface is maintained. If you find hot spots and wear in the round pipe after the bend, this too may need replacement. Replacing patched elbows with new wearbacks in one system reduced breakage and generation of fines from almost 5% to 1.7%.

In contrast to long-radius, rectangular section bends, Mark II elbows, develped by Rader Companies, Inc., are designed to reduce friction (power consumption) and protect chip fractionation. They are used at all points where bends of greater than 15° are required. Their design incorporates a smooth-flow elbow that projects into the central portion of a shrouded straight-flow pipe on the downstream side. The several components of the elbows have flanged ends for ease of removal and maintenance.

Design factors needed to determine power and air requirements for low-pressure pneumatic systems conveying wood chips are indicated in Table 5.1. For design of chip systems, the following two formulas should be used with these design factors:

$$scfm = saturation \times conveying\ rate\ (lb/min)$$

$$Pressure\ (psig) = hp/ton \times pressure\ factor$$

Velocity at the feeder may vary from 82 to 90 ft/sec. Some of these factors have been determined from actual installations; others have been determined by interpolation. As previously mentioned, the experience factor, so difficult to determine, should be given much consideration in the design of

TABLE 5.1. Design Factors for Pressure-Type Wood Chip Systems (capacity based on 210 cu ft of chips per cord at 25 lb/cu ft)

Capacity			Conveying Distance (ft)							
Tons per hour	Cords per hour	Factor	500	600	750	1000	1500	2000	2500	3000
52	20	Saturation:	1.8	2.2	2.6	3.2	3.8	4.5	5.9	7.5
		Hp/ton:	1.9	2.0	2.4	3.3	4.8	6.8	8.7	9.6
		Pressure factor:	3.3	3.0	2.5	1.8	1.3	1.0	0.8	0.7
78	30	Saturation:	1.7	1.9	2.2	2.6	3.2	3.9	5.0	6.1
		Hp/ton:	1.8	2.0	2.2	3.2	3.9	4.5	6.4	8.8
		Pressure factor:	3.2	3.0	2.7	1.9	1.6	1.3	1.1	0.8
104	40	Saturation:	1.6	1.7	1.9	2.3	3.0	3.7	4.8	5.6
		Hp/ton:	1.7	1.9	2.4	2.9	3.4	3.8	5.8	8.0
		Pressure factor:	3.6	3.1	2.5	2.1	1.8	1.6	1.2	0.9
130	50	Saturation:	1.5	1.8	1.9	2.3	3.0	3.7	4.5	5.4
		Hp/ton:	1.6	1.8	2.5	2.7	3.4	3.8	4.7	7.0
		Pressure factor:	3.9	3.5	2.4	2.2	1.7	1.6	1.5	1.0
156	60	Saturation:	1.5	1.8	1.9	2.3	3.0	3.7	4.5	5.2
		Hp/ton:	1.6	1.7	2.3	3.0	3.3	4.2	4.8	5.8
		Pressure factor:	3.9	3.5	2.6	2.3	2.0	1.6	1.5	1.1
195	75	Saturation:	1.4	1.7	1.8	2.2	2.9	3.6	4.4	5.1
		Hp/ton:	1.5	1.6	2.2	2.7	3.2	4.0	4.8	5.8
		Pressure factor:	4.0	3.6	2.8	2.4	2.1	1.8	1.5	1.1

such high-capacity and long-distance systems. Further, the differences in net area between successive pipeline sizes in large systems increases with each larger size. A borderline case in pipeline size determination requires a high degree of expertise.

In using these design factors, the operating pressure of all systems is between 6.0 and 7.0 psig. At these operating pressures, we can conclude that the systems can be classed as lightly loaded and can probably assume quite a bit of random movement of the chips within the pipeline. These pressures were no doubt arrived at by the desire to minimize blowback through the rotary feeders and to maintain an average velocity through the pipeline system. With chip breakage and fractionation a cause of great concern, would it not be advantageous to design the system for a pressure of 10 psig and reduce the saturation by possibly 20–25%? This would certainly give a more dense conveying stream, would reduce random movement, and might even decrease internal wear. While pipeline sizes would be reduced, power requirements would remain the same. Research in this vein could give us some interesting answers.

When conveying wood chips into bins, it is much better to separate the chips from the airstream in a cyclone receiver than to blow them directly into the bin. When they are blown directly into the bin, they compact and more readily interlock, causing difficulty in discharging out of the bin.

For screening out chip fines, dust, and dirt at the terminal end of the pneumatic system, a cyclone with a perforated plate inner cyclone for collecting the finer particles will separate them from the main flow of chips.

From the woodyard and woodroom operation, two other materials result: hog fuel (bark) from the barking drums and sawdust from the chip screens. From the barking drums, the bark is dewatered and collected by mechanical conveyors that deliver the bark to a hog. The hog breaks down the pieces of bark to a maximum dimension of 4 in. It is now called hog fuel and is transferred to the boiler house for burning. This transfer is adequately accomplished by a low-pressure system.

Conveying rates for hog fuel systems must be carefully determined. The conveying rate based on volume can be quite stable, but with wide variations in moisture the range by weight can vary considerably. A typical hardwood bark will weigh 24 lb/cu ft at 42% moisture, or 13.2 lb/cu ft on a bone dry basis. When the average moisture content is 34%, the bark will weigh 20 lb/cu ft, and at the maximum moisture content of 50%, the bark will weigh 26.4 lb/cu ft. Transposing this into tons to be handled per 24-hour day for a medium-sized mill, at 34% moisture 216 tons will have to be conveyed, and at 50% moisture the quantity to be conveyed will be 285 tons. If the conveying system is to convey directly from the discharge of the hog, the conveying equipment should handle 25% more than the hog will produce.

The feeding of hog fuel to the conveying system should be metered, if at all possible. Surges should be leveled out prior to entry into the rotary feeder. The feeder itself should have a volumetric capacity of about one-third greater than actual requirement. The conveying pipeline should have an inside diameter of at least twice the largest dimension of the hog fuel being conveyed. To withstand the abrasive action of sand and dirt in the bark, the pipeline should be of heavy construction and should include rectangular section bends with renewable wear plates at points of directional change. Although the wood is washed before and during the barking operation, the sand and dirt imbedded in the bark cannot be completely removed. The cyclone receiver at the terminal end of the system is also subject to extreme wear, so it, too, should be heavily constructed and equipped with abrasion-resisting liners at points of greatest wear. The cubic feet of air required to convey a pound of hog fuel (saturation) should be about twice that used for wood chips.

For conveying of the sawdust and wood fines from the screens, the low-pressure conveying system is adequate. Such systems, however, should use a maximum operating pressure of possibly no greater than 4 psig. This will minimize blowback through the rotary feeder used to charge the sawdust into the conveying pipeline.

With mills now making pulp from sawdust and shavings through M & D continuous digesters, sawdust is now becoming a vital part in the overall wood utilization program. Sawmills that previously burned their sawdust now ship it to the pulp mill. In transferring the sawdust from conveyances to storage silos, and from silos to surge bin above the digester, the pneumatic conveyor should be considered. It can furnish this transportation in an enclosed atmosphere, void of extraneous structures, and very economically, not only from an original equipment cost but also from a low operating cost and with minimum maintenance.

Sawdust systems can be of standard construction, with the conveying pipeline being either of welded steel or standard schedule 40 steel pipe. Pipeline bends should be reinforced against wear. Conveying velocity should be about 83 ft/sec at the feeder. Feeders for use in low-pressure systems can be the plain drop-out type, the blow-through, or the combination blow-through, drop-out. If combination vacuum–pressure systems are used, two rotary feeders should be installed, one at the discharge of the cyclone-receiver, then one directly below it to feed the sawdust into the pressure pipeline. Two feeders are necessary to minimize the differential across the feeders. This reduces blowback, which causes abnormal separation in the cyclone receiver, and holdup of the sawdust in the receiver cone.

A future material is waste wood, provided present feasibility studies prove that receiving tree-length wood with limbs attached for obtaining both pulp chips and waste wood for fuel is economically sound. Waste wood, if present research proves successful, may be added, together with hog fuel, to the lime sludge feed of the lime kiln in the chemical recovery area. This would reduce the amount of gas or oil required to fire the kiln. Another use could be as fuel for boilers. Its fineness for good combustion is said to be 100% below 10-mesh, with 75% below 18-mesh. Considering particle sizes and distances to be traversed between the woodyard and the points of use, pneumatic conveyors should be considered for both operations.

At this point in the pulp-making process, the wood, either chips or sawdust, is cooked in the digesters to make pulp. After cooking, the pulp is washed to remove cooking liquors and passed over screens or knotters to separate overize material such as partially cooked pulp, knots, and slivers. With the ever-increasing size of pulp mills, the resultant knots, slivers, and pulp from the screens are of sufficient quantity to warrant their being returned to the digesters for recooking.

The low-pressure pneumatic conveying system is the most logical type for this operation. It is not, however, a simple solution. Much coordination is required in the choice and operation of the screening equipment and collecting conveyors prior to entry to the pneumatic conveying systems. Furthermore, the material to be conveyed will be wet and moist, containing surface and impregnated cooking liquor.

Evidence of this is a design basis for such a pulp knot reject system for a 350-ton/day bleached pulp production mill. We can estimate that the normal rate of rejected pulp from the vibratory screens is 4% of the bleached production, or 14 tons/day during periods of normal production. Further, in periods of increased production, the rejects may increase 20%, for a total of 16.80 tons/day. To allow for occasional raw cooks, the design rate of the conveying system will be established at 20 tons/day at a consistency of 40% air dry.

Transposing this to specific figures, we find the quantity of dry solids in the hardwood and sulfite knots to be 1500 lb/hr (bone dry weight), the quantity of liquor, 2250 lb/hr, and, totaling these, we find a total quantity of material of 3750 lb/hr, weighing 46 lb/cu ft. The minimum conveying rate should be 4000 lb/hr.

Most systems in this service use this minimum conveying rate of 4000 lb/hr. Whether the system is 150 or 400 ft long, 4-in.-diameter conveying pipeline should be used. Any smaller size will invite trouble from oversize knots and slivers. The velocity at the feeder should be a minimum of 100 ft/sec, and while we may determine that the operating pressure will be 6 psig, the blower should be powered to accommodate a $33\frac{1}{3}$% overload, or a total pressure of 8 psig. Raw cooked pulp requires an awful lot of effort to keep it moving through the pipeline system.

Rotary feeders for feeding these rejects into the conveying pipeline should be of heavy-duty construction. They should be equipped with air-purged lantern rings in the stuffing boxes of the shaft seals to prevent the liquor from entering the seals, and a shearing bar at the throat to shear any particles caught between the rotor and the body at that point. The feeder should be direct-connected to its driving mechanism to avoid distortion and to have a minimum volumetric capacity of three times the volume of material being passed through it.

At the terminal end of the system, two methods of separating the rejects from the airstream can be employed. One is to deliver to a cyclone-receiver which in turn will discharge the rejects to a distributing belt conveyor over the chip bins. The other is to blow the rejects directly into the chip bins. The bins, in this case, are equipped with vent stacks to allow the conveying air to be vented at atompshere. After discharging these rejects into chip bins, they should be distributed among the chip supply to the digesters in order to achieve maximum results in recooking these rejects. They should not be allowed to accumulate in one lump, since they will adhere together, enter the digester en masse, and will again receive only a partial cook.

If a cyclone-receiver is used, it must be a high-efficiency type, having a minimum bottom cone angle of 75°. Further, the interior of the cyclone-receiver should have smooth surfaces so that the partially cooked pulp will not adhere to them. The bottom outlet should be of sufficient size to give a free fall to the material separated in the cyclone.

From the pulping operation in kraft mills, the spent sulfate cooking (black) liquor is concentrated and then burned in a recovery furnace as one step in the chemical recovery process. To replace the sodium salt and sulfur lost in the cooking process, salt cake (sodium sulfate), in the average amount of 100 lb for each ton of pulp produced, is added to the concentrated black liquor prior to burning in the furnace.

With the minimum economic plant size in many parts of North America reaching 500 tons of pulp per day, it is easily conceived that minimum requirements of salt cake usage will be 50,000 lb/day. Larger mills of 1200–1500 tons indicate the need for extensive handling and storage facilities to meet their daily salt cake demand. Pneumatic conveyors of the vacuum type are used to unload much of this salt cake from railroad boxcars and covered hopper cars, and deliver to vertical steel storage silos or bins. To reclaim the salt cake from the storage silos or bins, and deliver to the service bin in the recovery unit, both vacuum and low-pressure systems are used.

Salt cake is a material of varying physical characteristics. Its bulk density in the poured state varies from 70 to 108 lb/cu ft; in the tapped state from 79 to 109 lb/cu ft; and in storage bins from 70 to 108 lb/cu ft. The pH varies from 3.2 to 9.1; fineness from extremely fine (pulverized) to coarse granular. The angle of repose varies from 30 to 55°, which dictates that gravity spouts should have a minimum slope of 55° from the horizontal and cone bottom storage bins should have an angle of 60° to assure discharge flowability under any conditions and type of salt cake stored. The angle of repose of salt cake is very susceptible to moisture; it is hygroscopic, but not deliquescent. This characteristic can cause crusting of the top surface of a pile in flat storage. In fact, if allowed to remain in such storage long enough, dynamite might be needed to loosen it up. In vertical storage bins, no perceptible trouble is encountered provided the material in the bin is kept live by periodic withdrawal. If two or more vertical bins are used in the same installation, care should be exercised in scheduling withdrawals from the bins so that material movement out of, and within the bin occurs at least once every 48–72 hr.

Some varieties of salt cake can be classed as free flowing. Others can be extremely sluggish in flow, so much so that the material will stand up in a straight wall. Evidence of crusting and variable flow characteristics can be gleaned from excerpts of correspondence with a stevedoring contractor unloading salt cake from ships by a grab bucket.

In regard to crusting in the ships, we inspected a shipment quite closely, and found that in four of the hatches there was a negligible crust, easily penetrated with your hand. However, in one hatch there seemed to be a rather thick crust of about four inches. There is some speculation as to the cause of this crust as to whether it was caused from seawater entering the hatch, or from the hygroscopic nature of the material. In regards to the free flowing qualities of the salt cake, we noted it was not free flowing, and that a grab bucket could dig right down leaving a vertical wall. We have since inspected the material stored in a silo. It has to be handled physically to make it flow freely in the silo.

Another cargo of salt cake was received from a different manufacturer in Germany, and seemed to be a completely different product. This salt cake seems to be much heavier, more granular, and free flowing. It does, however, stand up in a solid wall when a grab bucket is digging into it. There was very little lumpiness in the cargo, and there was no surface crust whatsoever.

Variations in salt cake will affect conveying rates. One vacuum system unloading from cars and delivering to storage reported variations from 7 to 18 tons/hr. Another reported variations from 20 to 35 tons/hr. Salt cake systems should be designed to handle all types. From experiences of the past, in designing and installing any type of material handling equipment to handle salt cake, it is unwise to be conservative. Power requirements are high, and heavy construction of all equipment should be a prime requisite.

In the handling and storage of salt cake in the pulp mill, six arrangements of equipment are possible to unload the salt cake, store it, and deliver to its use point, the mixing tank for the recovery unit. Incidental to, and at times required when synthetic salt cakes are handled, a swing hammer pulverizer must be inserted in the handling circuit to break down the lumps to a $-\frac{1}{8}$-in. product. This is necessary to avoid fouling of the screens in the mixing tank, and the clogging of the spray nozzles of the recovery boiler. Some natural or highly refined salt cakes do not lump in transit or in storage bins so the pulverizer may not be necessary. It is good insurance, however, for even if a pulverizer is not immediately required and installed, space should be allowed for installation at some future time when pulverization is required. No mill will tie itself to a single supplier for obvious reasons. If the highly manufactured type is purchased today, it could be the synthetic type tomorrow.

Systems in use are vacuum unloading and pressure reclaiming using a single exhauster-blower to operate both legs not simultaneously; vacuum unloading and pressure reclaiming using separate exhausters and blowers so that both operations can be done simultaneously; vacuum unloading and vacuum reclaiming using a single exhauster to operate both legs, again not simultaneously; vacuum unloading and vacuum reclaiming using separate exhausters for simultaneous operation; vacuum unloading and low-pressure continual feeding to the mixing tank; and vacuum unloading and mechanical mixing system below or adjacent to the main storage bin.

A combination vacuum unloading and low-pressure reclaiming system using a single exhauster-blower (Figure 5.14) is satisfactory for a mill having a production of 600–700 tons of pulp per day. The usage of makeup salt cake of 125 lb/ton of pulp produced results in a daily total usage of 85,000 lb. When unloading either 50- or 70-ton cars, the mill will find sufficient time for the separate unloading and reclaiming operations to satisfy their requirements when conveying at the rate of 10 tons/hr. In this system, the vacuum leg is simple and straightforward. The reclaiming leg, which can be operated only when the vacuum leg is shut off, requires special consideration in order to minimize dusting from the pulverizer. All of the air to the intake of the exhauster-blower is drawn in through the dust line and bleeder intake above the pulverizer with the dust from the pulverizer being retained and discharged by the filter-receiver into the storage bin. The salt cake after passing through the pulverizer is charged into the low-pressure reclaiming pipeline through the line charger rotary feeder. The pipeline terminates in a single-stage cyclone-receiver in the recovery building. This receiver, in turn, discharges into a surge or service bin equipped with a high level bin indicator. When material reaches this bin level indicator, the vibrating pan feeder below the storage bin is shut down, and the system allowed to operate for a few minutes for the system as well as the pulverizer to clean out. The discharge air from the receiver is piped over to the boiler breeching, prior to entry into an electrostatic precipitator for the collection of dust that has emitted from the cyclone-receiver with the air. For a 10-ton conveying rate, and the exhauster-blower sized for vacuum service, the amount of air delivered to the precipitator is approximately 1800 cfm. This additional air has no effect on normal operation of the boiler and its appurtenances. In designing such a system, the filter-receiver should have sufficient filter media to accommodate the amount of air passing through it under reclaiming conditions.

When the pulp mill reaches a production rate of 1000 tons/day, it is advisable to divorce the unloading and reclaiming systems into separate operations (Figure 5.15). In this type of installation, salt cake is unloaded from either of two car unloading positions to the storage bin through a simple vacuum system. The second car unloading position obviates the need to move the cars in the event two cars are received and only a single unloading station is provided. For delivery of the salt cake to the service bins

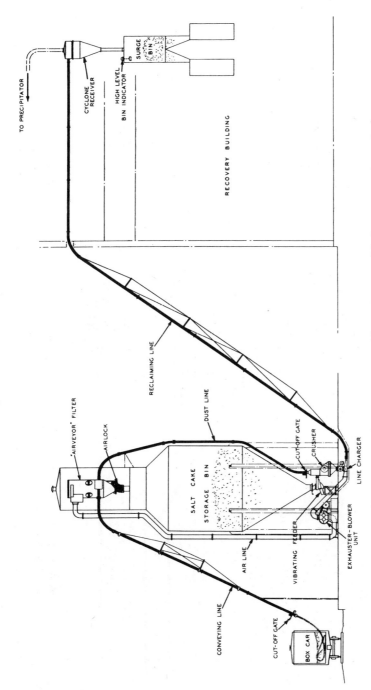

FIGURE 5.14. Combination vacuum unloading and low-pressure reclaiming system using a single blower. (Courtesy *Tappi.*)

195

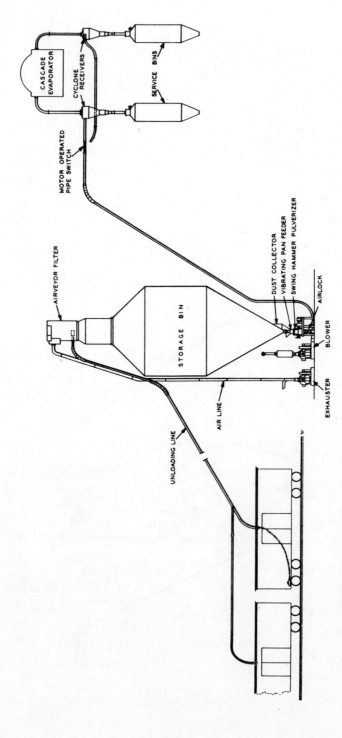

FIGURE 5.15. Separate vacuum unloading and low-pressure reclaiming system using separate blowers. (Courtesy *Tappi.*)

in the recovery unit, a low-pressure system is used with a motor operated pipe switch or diverting valve in the conveying pipeline to divert the salt cake to either of the two service bins. We should note that when the systems are divorced, the filter-receiver cannot be used for dedusting the pulverizer. For this purpose, a small dust collector is placed above the junction hopper between the vibrating pan feeder and the swing hammer pulverizer. Like all equipment handling salt cake, this dust collector should be of heavy construction and should have a good factor of safety regarding its cloth area.

The air discharge from the cyclone receivers terminates in the Cascade evaporator of the recovery unit. The entrance of 1290 cfm of air from the separate pressure system has no adverse effect on the operation of the evaporator. By discharging the air and dust into the Cascade evaporator, the collected dust is mixed with the black liquor before entry of the liquor into the mixing tank. This reduces the dust load on the electrostatic precipitator.

In some cases, mill operators and engineers prefer a complete vacuum handling system so that the evaporator, ID fan of the boiler, and the precipitator receives no external air and dust other than that from the boiler itself. A mill of 600 tons/day production installed such a system (Figure 5.16). This installation uses the vacuum system either to unload the cars or to reclaim from below the pulverizer and deliver to a single-stage cyclone receiver in the recovery unit. The air line from the cyclone receiver connects to the filter-receiver above the storage bin. Rotary feeders or discharge locks are required below both the filter-receiver and the cyclone receiver. The system uses only one exhauster unit which operates only under vacuum service.

It will be noted that with a vacuum pickup below the pulverizer, no additional equipment is required to dedust the pulverizer. Part of the conveying air entering the conveying pipeline is brought through the pulverizer, causing a slight negative pressure to be placed on the unit, as well as drawing air through, which not only aids in the operation of the pulverizer, but avoids the emission of dust between the shafts and the housing.

The filter-receiver in this all-vacuum type system will have to retain the dust from the two operations, and should be sized accordingly. While the operations cannot be carried on simultaneously, the additional operating time on the filter-receiver should receive consideration.

A pulp mill expanding to 1400 tons/day production required conveying equipment to replace existing and somewhat antiquated mechanical conveyors to serve their existing recovery units, plus a large new unit needed with the expansion in pulp production. A system as dust-free as possible was desired. Requirements called for an unloading rate of 25 tons/hour, and a reclaiming rate of 20 tons/hour. To adequately satisfy these requirements, two pneumatic conveying systems were installed (Figure 5.17).

To unload from the cars and deliver to the storage bins, a vacuum type unloading system actuated by a 150 hp exhauster unit was used. The dis-

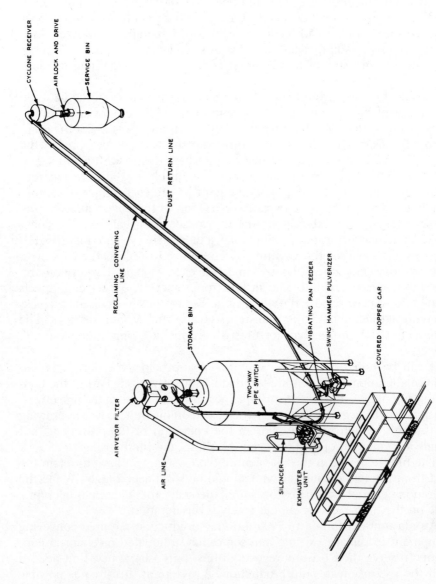

FIGURE 5.16. Combination all-vacuum unloading and reclaiming salt cake system. (Courtesy *Tappi.*)

CYCLONE RECEIVER

AIRLOCK AND DRIVE

SERVICE BIN

DUST RETURN LINE

RECLAIMING CONVEYING LINE

STORAGE BIN

VIBRATING PAN FEEDER

SWING HAMMER PULVERIZER

TWO-WAY PIPE SWITCH

AIRVEYOR FILTER

AIR LINE

SILENCER

EXHAUSTER UNIT

COVERED HOPPER CAR

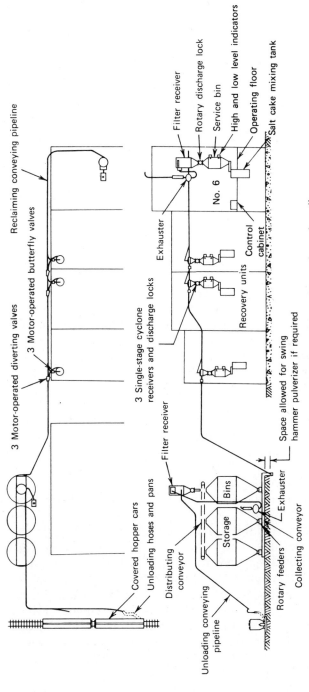

FIGURE 5.17. Salt cake systems for 1400 ton/day pulp mill.

Reclaiming conveying pipeline

3 Motor-operated diverting valves

3 Motor-operated butterfly valves

Filter receiver

Rotary discharge lock

Service bin

High and low level indicators

Operating floor

Salt cake mixing tank

No. 6

Exhauster

3 Single-stage cyclone receivers and discharge locks

Recovery units

Control cabinet

Space allowed for swing hammer pulverizer if required

Covered hopper cars

Unloading hoses and pans

Filter receiver

Distributing conveyor

Storage

Bins

Exhauster

Unloading conveying pipeline

Rotary feeders

Collecting conveyor

charge from the filter-receiver was delivered to the three storage bins by an existing drag-type mechanical conveyor. For reclaiming to the recovery units, it was desired to deliver the salt cake to the service bins of four recovery units by a vacuum system. While we would normally think of a low-pressure system when one pickup point and a multiplicity of discharge points are served, the choice of a vacuum type system would be quite a radical departure. In a way, it is. But if any leakage occurs in the conveying equipment, the leakages are in, not out, so that a dust problem from this source in the recovery units would be nonexistent. For such requirements, vacuum systems do have advantages.

The vacuum-type reclaiming system is also powered by a 150 hp exhauster unit. Its conveying operation begins at the discharge of a collecting drag conveyor below the storage bins. A pickup hopper with a stationary type nozzle, pipeline, intermediate cyclone receivers, and a terminal filter-receiver serves to convey the salt cake to any one of the recovery units. Through a control panel in recovery no. 6, together with high and low level indicators in the service bins, and automatic switches in the air and conveying pipelines, the operators can man the operation from a single operating position. While a pulverizer was not installed, space was allowed for it between the discharge of drag conveyor below the storage bins and the pickup to the conveying system. In the event that a pulverizer becomes necessary, it can be installed at this point with minimum change to the system as installed.

Another point in separating the unloading and reclaiming systems is that in many mills, the yard gang has the responsibility of unloading the cars, and the operating people have responsibility of seeing to it that salt cake is always available for feeding into the mixing tank. By having two separate systems, the cleavage between the areas of responsibility is clean cut, avoiding overlapping and operational jurisdiction.

A desire to eliminate the service bin in the recovery unit resulted in the application of a low-capacity, continuous, low-pressure system that delivered salt cake from the discharge of the storage bin directly to the mixing tank (Figure 5.18).

The application of this arrangement is a mill having a production of 300 tons of pulp per day. The makeup of 125 lb of salt cake per ton of pulp produced, resulted in a daily usage of 37,500 lb/day. For unloading the salt cake, a vacuum system of 8–10 tons/hr conveying rate was installed. The separate low-pressure-type system was sized to convey salt cake at the rate of 6000 lb/hr to take care of any abnormal requirements of the mixing tank. The system at the discharge of the storage bin uses a variable speed, gravimetric type feeder to regulate the feed of salt cake to the swing hammer pulverizer and conveying system. Below the pulverizer, a rotary feeder charges the salt cake into the conveying pipeline, which terminates in a single-stage cyclone receiver above the mixing tank. The discharge air from the cyclone receiver is passed into the Cascade evaporator.

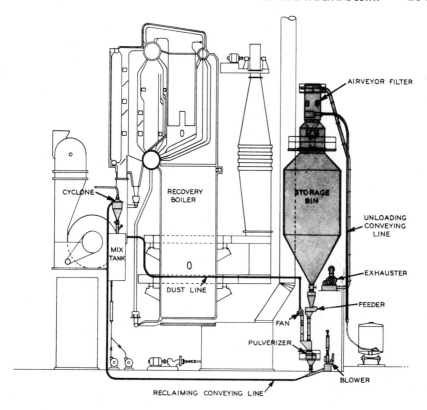

FIGURE 5.18. Separate vacuum unloading and continuous low-pressure salt cake reclaiming system. (Courtesy *Tappi.*)

When the system was first installed, a standard bin vent dust collector, actuated by a $\frac{1}{2}$-hp fan was used to maintain a slight negative pressure in the pulverizer, together with an airflow to maintain dedusting. This bin vent type of dust collector, being of standard construction, was not sufficiently strong to stand continuous operation and the extremely hard service. It was replaced by the fan system, which proved to be very successful.

Some mill operators and engineers went one step farther, and chose to move the mixing tank from the recovery unit down to grade, either directly below, or adjacent to the storage bin (Figure 5.19). This system again uses the straight vacuum type system for unloading from the cars and delivering to the storage bin. For mixing the makeup salt cake with the black liquor from the evaporators and wet bottom precipitator, the salt cake is discharged from the storage bin through a vibrating pan feeder, and then passed through a swing hammer pulverizer to a sloping screw conveyor. This sloping screw conveyor delivers the salt cake to a Hummer screen to separate oversize lumps of salt cake and other materials not broken down by the pulverizer. The screened material discharges into the mixing tank. The black liquor from the mixing tank is pumped in slurry form to the Cascade

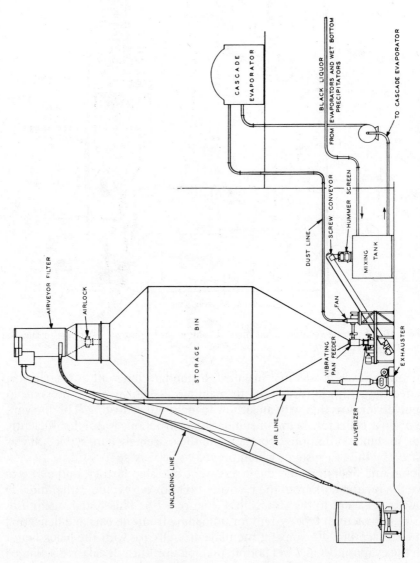

FIGURE 5.19. Vacuum unloading with salt cake mixing tank at grade. (Courtesy *Tappi.*)

evaporator. To dedust the pulverizer, the fan system was again used. The pulverizer is placed under the slight negative pressure by the vacuum side of the fan, the air and salt cake dust from the pulverizer is passed through the fan, and then by pressure from the fan is blown through the dust line to the Cascade evaporator. To minimize frictional resistance in this dust pipeline, cold drawn tubing should be used.

When pulverizers are used in conjunction with low-pressure reclaiming systems, their dedusting should be controlled. In most cases, a 3-ton/hr capacity swing hammer pulverizer will require 405 cfm for dedusting; a 10-ton/hr pulverizer, 850 cfm; a 15-ton/hr pulverizer, 1450 cfm; a 20-ton/hr pulverizer, 1700 cfm; and a 30-ton pulverizer, 2525 cfm, all at 2 in. water gauge vacuum.

In discharging salt cake from vertical storage bins, vibrating pan feeders generally are the most applicable. When handling the highly manufactured type of salt cake that remains relatively free flowing, rotary feeders can be used to feed the salt cake out of the storage bin into a vacuum conveying line. These feeders, however, should have the clearances between the rotor and the body widened to 0.015 in. When discharging into a low-pressure system, a mechanical feeder, preferably the vibrating pan, should be used to feed from the bin outlet to the rotary feeder charging the conveying pipeline. This feeder, however, should have normal clearances. Above the inlet to the rotary feeder, a junction hopper with some type of air vent line is needed to allow the blowback air from the feeder to dissipate to either atmosphere or dust collecting equipment. The rotary feeder should be equipped with air purge lantern rings in the shaft seals to minimize maintenance on the packings.

Among other losses of chemicals in the kraft pulping operation, lime, in the amount of 25-lb/ton of pulp produced, is lost. To take care of this loss, fresh lime is added to the reburned lime discharged from the rotary kiln of the chemical recovery plant. The lime is usually received by the mill in covered hopper cars. The unloading of this lime from cars and delivering to the fresh lime bin is an excellent application for a vacuum type system. Such an installation (Figure 5.20) serves two causticizing areas. The lime is unloaded from either of two car-unloading stations and delivered to either of two fresh lime bins over conveying distances of 110 and 630 ft, utilizing a single exhauster unit.

Smaller mills can take advantage of pneumatic conveying systems through their ability to unload both salt cake and lime through the same system (Figure 5.21). In such a system, the salt cake and pebble lime should be separated at the outlet of the filter-receiver. The salt cake is discharged through one leg of a two-way gate directly to the salt cake storage bin. The lime is delivered to the horizontal screw conveyor through the other leg of the two-way gate for delivery to the pebble lime bin. In this instance, the salt cake being handled was limited to the highly manufactured type since a rotary feeder is attached directly to the outlet flange of the storage bin.

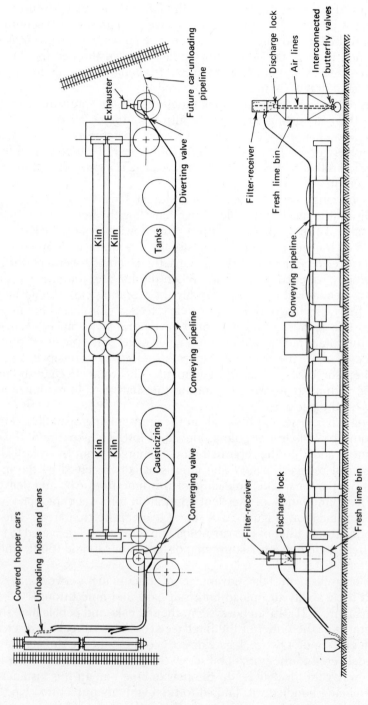

FIGURE 5.20. Lime unloading system for 1400 ton/day pulp mill.

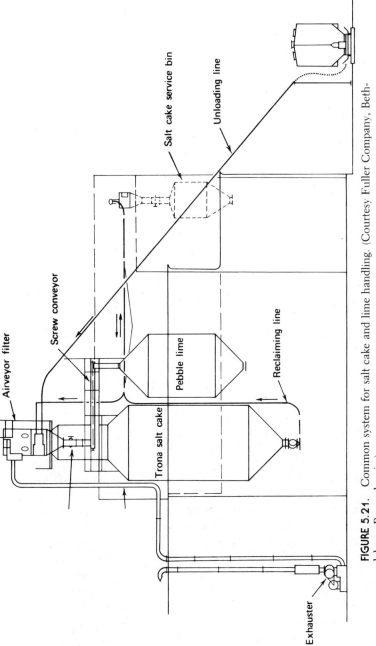

FIGURE 5.21. Common system for salt cake and lime handling. (Courtesy Fuller Company, Bethlehem, Pennsylvania.)

From this rotary feeder, the salt cake is conveyed by vacuum to the service bin in the recovery unit. The lime from the fresh lime bin is delivered to the slaker through a weightometer as required.

In some mills, the lime mud from the vacuum washer of the chemical recovery system requires the addition of limestone to the feed to the rotary kiln for calcining. The vacuum type unloading system can be very effectively applied in the conveying of this usually sized $\frac{3}{8}$ in. and under limestone from the cars and deliver to the kiln feed bin. This limestone is unburned lime, and abrasive. A normal schedule no. 40 conveying pipeline is used with reinforced bends. At the discharge of the filter-receiver, gate locks should be used since the abrasive nature of the limestone would wash out a rotary feeder in a relatively short time.

Pneumatic conveyors should *not* be used for conveying hot lime as it is discharged from the kiln. Mechanical conveyors such as a drag for horizontal conveying and a bucket elevator for elevating should be used. These units must be specially built to withstand the heat from the lime as well as abrasion.

About 10 years ago, a hot lime system was installed to convey from the discharge of a rotary kiln 500 ft to a cyclone above a hot lime bin. With much fanfare, the system was discribed as the only system in North America where hot lime discharged from the kiln was conveyed by a low-pressure pneumatic conveying system. Downtime and high maintenance were of such magnitude that about a year after startup, a slaker was installed at the hot lime end of the kiln, replacing the pneumatic system, which proved unsatisfactory. Mechanical conveyors were installed to convey the lime from the kiln to the slaker.

In the soda pulp mill, the materials that lend themselves to the application of pneumatic conveyors are soda ash and pebble lime. The vacuum system is applied for unloading these materials from railcars and delivering to their respective storage bins. In making up the cooking liquor in the mill, the materials are mostly metered directly to mixing equipment from the storage bins.

In the mechanical pulp or groundwood mill, pneumatic conveyors can be adequately applied to the conveying of dried or pelletized pulp, through either vacuum or low-pressure systems. The low-pressure system can blow the dried pulp into boxcars for delivery to a converting mill, where the same pulp can be unloaded by a vacuum system, from the cars to storage. In the handling of dried pulp, cyclone receivers only are required at the terminal end of vacuum systems, but there should be two stages, sufficient to effect separation of air and material so that only air passes through the exhauster.

In the conveyable state, so-called dried groundwood will have a bulk density of approximately 15 lb/cu ft and a moisture content of around 40%.

The neutral sulfite, semichemical pulp mill requires two materials in sufficient volume to merit bulk handling; sodium sulfite and soda ash. Both of these materials are received in railcars; the sodium sulfite in either covered hopper or boxcars depending on flowability, and the soda ash in covered hopper cars. The unloading of cars and conveying to storage merits the consideration of the vacuum-type system.

Sodium sulfite is a material quite similar to salt cake regarding flowability and bulk density. It is also hygroscopic. Its bulk density will vary from 75 to 97 lb/cu ft, and for satisfactory conveying rates, its moisture content should not exceed 1.25%. For a vertical conveying distance of 75 ft and minimum horizontal run, the saturation (cubic feet of air per pound of material conveyed per minute) is 2.5. Add to this a horizontal run of 75 ft, the saturation required then is 3.4. While this material is very erosive when entrained in air, black steel construction has appeared to have a reasonable life-span. Flexible metal hoses connecting the car-unloading hoppers to the conveying pipeline have shown extreme wear in carbon steel construction. The substitution of an all stainless steel hose will to a great extent, alleviate this problem. Air-to-cloth ratios for filter-receivers are the same as those required for salt cake.

The sodium sulfite is stored in the dry state. The soda ash, however, is stored in slurry form in most mills. In conveying both of these materials through the same vacuum-type system, the soda ash will present no problems. The sodium sulfite, delivered in covered hopper cars, will require vibration to maintain flow out of the cars. It will also have varying conveying rates, depending on its physical characteristics.

The deinking mill has limited application to the use of pneumatic conveyors. Where the deinking pulp mill and paper mill are one homogenous process, the only application is the transfer of rejects from dewatering equipment to either boilers for burning or conveyances to land fill operations. In this application, the amount of dewatering affects the conveying system. The dewatering equipment should be capable, at all times, of delivering the pulp rejects to a low-pressure type system at a consistency of 40% (combined moisture, 60%, no free moisture). Any rejects having a lower consistency will be extremely wet and difficult to convey. A low-pressure system for this service should be designed to have a minimum velocity at the pickup point below the charging rotary feeder of 90 ft/sec. All pipe joints are to be smooth inside so that buildup at these points in the conveying pipeline is avoided. The rotary feeder, sized to act as an airlock and not as a feeder, should have protective devices to prevent clipping between the rotor and the body, and to be direct connected to its driving mechanisms to prevent distortion.

When the deinking pulp mill is separated from the paper mill by more than a few miles, the deinked stock must be transported, either by railcar or van type truck. To eliminate paying freight charges on water, the stock is dewatered by a press to 45–50% consistency. In this state, the pulp is in nodule form and of fairly stable particle size. From the press, the low-pressure system is an ideal application for loading the pulp into the railcars or trucks. By directing the flow of material into the cars or trucks, a degree of compaction results in the pulp attaining a bulk density of 22 lb/cu ft.

At the paper mill, a vacuum system capably unloads the pulp to storage, void of contamination from dirt or other material. Cyclone receivers having a minimum of two stages of separation are necessary to separate the pulp from the airstream. As the pulp is transferred through the conveying pipeline and separated in the cyclone receiver, it is subjected to an ever increasing negative pressure. This negative pressure explodes the pulp nodules to such an extent that their bulk density is reduced to 12–14 lb/cu ft. The discharge locks below the cyclone receiver are therefore sized for this bulk density, not the bulk density in the railcar or truck. While this reduction in bulk density requires increased storage volume to store the pulp, there is a distinct advantage to the exploded nodules. When the pulp is put into the hydra-pulpers and beaters, it is found that the time required for the beating operation was reduced some 50–60%.

So far, we have concerned ourselves with the application of pneumatic conveyors in the pulp mill. The paper mill has applications similar to those of the pulp mill. As pulp is refined, fillers like clay (loading), dye, size, alum, and starch are added to the stock prior to its formation into a sheet of paper. After the paper is formed on the paper machine, sizing (starch), clay (coating), and pigments are applied to the paper, giving it the proper gloss and finish for its end use. Materials used in sufficient quantity by most mills to warrant bulk handling and storage include clay, starch, and calcium carbonate.

To determine amounts of usage and conveying rates, both unloading and reclaiming, many variables are encountered that affect such rates. For general purposes, however, the following clay usage rates for different types of papers and coatings can be applied:

| | Pounds of Clay per Ton of Paper | |
Type of Paper Produced	Loading	Coating
Uncoated papers	160–400 lb	
Litho coated—one side	100 lb	485 lb
Letterpress coated—both sides	180 lb	485 lb
Magazine machine coated—both sides	225 lb	500 lb

The rate for starch usage is determined by the method of coating and the coating formulation. Since coating formulations differ considerably, we can again assume general purposes and the usage as follows:

Method of Coating	Pounds of Starch per Ton of Paper
Brush or air blade	75–125 lb
Machine	60–100 lb

Calcium carbonate usage may vary from 5–50% of the total coating formulation. A good average coating formulation of the three primary materials would be the following:

Predispersed clay	85 parts
Starch	15 parts
Carbonate	15 parts

The tons of paper produced per day multiplied by the usage of clay and starch per ton produced will give us the total usage per day. From this, receipts per week or month can be determined, and then the required conveying rate, depending on the desired hours of operation, can be found.

Papermaking clays are produced in various physical forms. These forms present various physical properties such as bulk density, moisture content, and flowability, which, when combined, affect conveyability. Typical characteristics of the various types and their conveyability are indicated in Table 5.2. These are average figures, and may vary from supplier to supplier.

In the handling of clay pneumatically, the conveyability is determined mostly by the moisture content. When the basic conveying rate is determined, it will be maintained as the moisture content varies from 0–3%. Above 3%, there will be a reduction in the conveying rate of 10–15% for each additional percent of moisture. A system designed to convey clay at 0–3% moisture at a rate of 10 tons/hr, will handle only 7–8 tons at 5% moisture content.

The variation in bulk density is affected not only by its physical form but also by aeration during conveying. As the clay is conveyed by the airstream through the pipeline, aeration makes it more fluid, reducing the bulk density as it is discharged from the conveying system. The bulk density of lump

TABLE 5.2. Clay Characteristics

	Bulk Density (wt per cu ft)	Moisture Content	Conveyability
Filler (loading) grades			
Air floated, pulverized	30	1–2%	Good
Water washed, lump	52–72	2–7%	Fair
Pulverized	21–40	0–1%	Good
Pulverized	35–40	2–4%	Fair
Pulverized	40–60	1–3%	Good
Coating grades			
Water washed, lump	44–69	2–7%	Fair
Pulverized	18–30	1–3%	Good
Pulverized	35–40	2–4%	Fair
Spray Dried	50–60	0–1%	Good

clay after conveying is reduced about 15–20%; pulverized clay by 20–35%; and spray-dried clay, as much as 40–50%. This reduction in bulk density and subsequent higher bulking calls for much care in sizing the rotary discharge lock below the filter-receiver of a vacuum system, as well as mechanical conveyors, weigh hoppers, and storage bins, downstream from the conveying operation. After delivery to the storage bin, clay will settle or deaerate. Its bulk density will then return to somewhere between 75 and 90% of the bulk density it had in the railcar before conveying. Some clays, notably the lump type, will deaerate in storage much more rapidly than others. The pulverized and spray-dried clays will take much longer.

We should, at this time, recognize the fact that spray-dried clay after being unloaded pneumatically is no longer in bead form. During its travel through the pipeline, the beads are broken with the result that we have a form that is very near to the pulverized state. This affects only storage and flowability thereafter, not the end use.

Clay is stored mainly in vertical storage bins in the dry state. Some flat storage is used to store some of the lump and high moisture clay since it is most difficult to get this type of clay out of vertical storage bins having cone bottoms. Pulverized clay, including spray-dried after conveying, can be stored easily in vertical storage bins. The discharge from these bins can be assisted through the use of aeration in the bottom cones. Reclaiming hoppers (Figure 5.22) are used in both flat- and cone-bottom bins to facilitate flow out of the bins. In the flat-bottom bins, aeration units are installed around the perimeter of the opening to aerate the clay adjacent to the opening. This aeration causes the clay to flow like water, minimizing the amount of dead storage within the flat-bottom bin. In the cone-bottom bin,

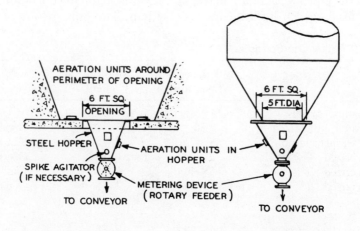

FLAT BOTTOM BINS   CONE BOTTOM BINS

FIGURE 5.22.   Clay discharge from bins and silos. (Courtesy *Tappi.*)

the break in dimensions of the round cone bottom and the rectangular reclaiming hopper tends to break down arches of material between the outlet of the hopper and the cone bottom itself. With aeration units in the hopper, a good flow can be maintained out of the bin. It is also good insurance to have the bottom cone and hopper treated with a special smooth interior coating to facilitate flow with minimum resistance. This reduces the friction between the clay and the bin walls, inducing freer flow.

In handling and storing high-moisture clay, it has been found that when the bin storing this material is more than half full, that the moisture in the upper level dissipates from these levels to the clay at the bottom of the bin. This causes the moisture content at the bottom of the bin to be 20–40% higher than at the upper levels, which leads to withdrawal difficulties.

In recent years, the trend has been toward slurry rather than dry storage of clay. Slurry storage is said to be able to deliver a clay slip with minimum variations in solids content, minimum impurities, and maximum homogeneity. Slurry clay stores in less space, with advantage taken of the full volume of tanks. Slurry storage of clay at 70% solids will store almost $2\frac{1}{2}$ times that of pulverized or air floated clay, and 56% more than spray-dried.

When conveying clay and starch in the same system, a not too clearly defined phenomenon exists. This phenomenon, related to mineral (clay) and vegetable (starch) materials, illustrates the variance in conveyability between materials of such different classification. In a 5-in. vacuum system, conveying both clay and starch from covered hopper cars to a filter-receiver located about 50 ft above the car unloading station, the designed conveying rate of 8 tons of clay or 10 tons of starch per hour was used. Actually, the system had to unload starch at the rate of 12–14 tons/hr to achieve the 8-ton/hr rate on clay. In an 8-in. system, the conveying rates for both the clay and starch are about equal. Knowing when to apply proper conveying rate to systems conveying more than one material is the art in the design of pneumatic conveying systems. When designing systems to handle both clay and starch, base the design on the handling of clay. The materials can then be conveyed at or above the required rates.

Systems handling starch should be explosion proofed. All equipment should be grounded to dissipate static electricity. Conveying pipeline joints should have static jumper wires across them. The filter-receiver should be equipped with explosion vents, and the filter media with static wires. Before entry of the starch into the storage bins, magnetic separators should be used to collect tramp iron from the conveying stream.

A good illustration of the need for magnetic separation to separate nuts, bolts, nails, wire, etc., from the conveying stream beside explosion proofing is the problems encountered by a small New England mill. This mill received its starch in Airslide cars, and conveyed the starch from the cars to slurry through a small, low-pressure system. The mill suffered serious maintenance problems with the rotary feeders of the conveying system as well as the pumps for the starch slurry. No magnetic separation was included in the installation. The nuts, bolts, nails, and wires that would have been separated had this been included, caused all the havoc.

The mill, probably through ignorance, was blaming the carbuilder for all this trash in the starch. The car, however, was under a lease arrangement to the supplier, and had been traveling over the rails for eight months. The last time the carbuilder saw the car was when it left its shops after completion and lease to the supplier. If abnormal trash, and most important, tramp iron, is included with the starch, check with the supplier. He should and will eliminate the source when so advised.

It is good planning also to include magnetic separation when handling clay only, especially when unloading cars and delivering directly to mixers or mixing tanks for slurry storage. One high-cost maintenance job can very well outweigh the saving of cost of magnetic separation, to say nothing about the loss in production.

Historically, the pneumatic conveying of bulk clay was introduced to the paper industry during the depression years of the early 1930's by the clay producers. At that time, cost reduction was probably uppermost in the mind of every mill manager, but capital expenditures were just nonexistent. The clay producers, in order to effect sale of their product and promote bulk handling, financed the installation of the early pneumatic conveying systems. The mills paid off the equipment with the difference between the cost of bagged and bulk. They paid bagged prices for the bulk clay until the cost of the equipment was amortized, at which time the mill took complete ownership of his installation and the reduced rates for the clay. Bulk handling, by eliminating bagged handling and cumbersome unloading methods, enabled the mill to reduce the labor costs in handling clay and to amortize the conveying equipment at the same time.

The early installations handled clay from boxcars only. The vacuum system was employed to unload the car, and distributing screw conveyors delivered the clay to flat storage areas. In order to mix the clay for loading as well as coating, it had to be reclaimed from the flat storage area in the same manner in which the operator unloaded the clay from cars. The clay was sucked into the filter-receiver, discharged into the distributing conveyor, which, by reversing, delivered the clay to a weigh hopper for metering into the mixing tank.

In the early 1940s, clay usage had increased to the point where better handling and storage methods were needed. This brought forth the storage of clay in vertical storage bins constructed of steel, concrete stave, and tile. This type of storage was made possible by the newly acquired knowledge that aeration to the clay in the cone- and flat-bottom storage bins accelerated the flow out of the bins, avoiding costly hangups.

Batch mixing of the clay was still the accepted method, mixing dry clay with water and other pigments for the coating formulation. With inceased production requiring greater efficiency and faster formulation, slurry storage was perfected in the early 1950s. This allowed clay dispersions to be stored in slurry form and metered in such form to the formulation tanks. Further, with higher starch and carbonate usages, the pneumatic conveying systems are used to unload not only the clay, but these materials too. The system, being self-cleaning, is well suited to the handling of all three materials through the same system, maintaining minimum contamination.

An installation handling both clay and starch (Figure 5.23) indicates the flexibility allowed in pneumatic conveyng. Covered hopper cars of clay and Airslide cars of starch are unloaded from any one of three car unloading stations and delivered to their respective storage bins. This system is a straight vacuum type system, designed to convey the starch and clay at a rate of 30 tons/hr. The 8-in. conveying pipeline terminates in a filter-receiver above the storage bins. This filter-receiver is equipped with explosion vents and static dissipating wires in the filter media for protection when handling starch. Discharge from the filter-receiver is done through a straight-neck rotary discharge lock to maintain a free discharge of material from the filter-receiver cone into the pockets of the rotary lock. Below the discharge lock is an offset spout containing two permanent magnet type spout magnets for the removal of tramp iron. Delivery to the storage bins is through direct spouting and horizontal screw conveyors.

To maintain a supply of clay in the color building, a separate vacuum system having a conveying rate of 20 tons/hr is used. This system reclaims clay from the storage bins, also through an 8-in. pipeline. Aerated hoppers in the bottom of the flat slab of the storage bins provide a good flow of clay to the reclaiming system. From the filter-receiver at the terminal end of the system, the clay is deposited in the use bins by open-bottom horizontal screw conveyors. To feed the weigh hoppers for batch mixing, air-activated gravity conveyors are used to convey from the flat clay use bin to the hoppers. Around the inlet openings of the air-activated gravity conveyors, aeration units are used to fluidize the clay in close proximity to the inlets and to reduce the amount of dead storage.

For conveying the starch from the storage bins to the starch cooker, a low-pressure system is used. The starch is discharged from the storage bins by aeration hoppers and twin-screw feeders, to a rotary feeder which charges the starch into the conveying pipeline. The system terminates in a weigh hopper that is equipped with cloth filter tubes to retain the starch dust and, at the same time, to allow the conveying air to dissipate. This weigh hopper is a weighing out operation, providing accurate metering of dry starch for the cooking operation.

The filler (loading) clay is unloaded from cars through a separate $7\frac{1}{2}$ ton/ hr vacuum system. This system, in operation for many years, delivers the clay to a flat storage area adjacent to the car-unloading track.

In the unloading of clay from cars to slurry storage, while the vacuum-type system is used in both cases, the type and size of the mixing equipment determine the amount of equipment to be used in conjunction with the

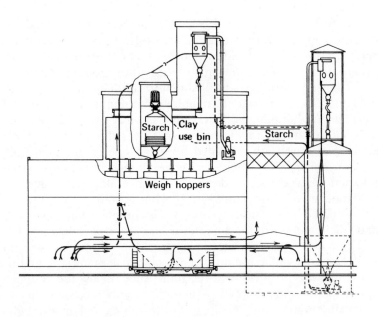

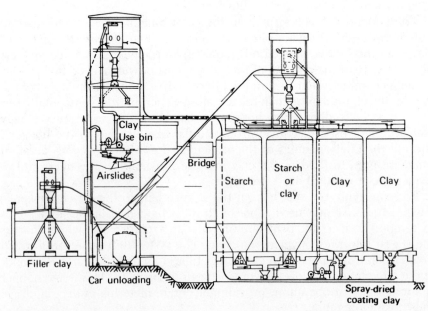

FIGURE 5.23. Clay and starch handling systems. (Courtesy Fuller Company, Bethlehem, Pennsylvania.)

conveying system. One method is to unload portions of the car into the mixing tanks, and the other is to unload and mix a full carload in a single tank.

The method of unloading part of the car and using the smaller mixing tank is indicated in Figure 5.24. In this installation, 50- or 70-ton capacity covered hopper cars containing spray-dried predispersed clay are unloaded by a vacuum system at a rate of 25 tons/hr. The mixing tanks were sized to mix clay and water in the amount of 25,000 lb of clay, mixed in the proportion of 10 lb of clay per gallon of slurry. To meter the clay into the mixing tank, an automatic scale is used having a capacity of 250 lb for each dump. Through a counter on the scale mechanism, after 200 dumps were metered through the scale hopper, the flow of clay through the conveying pipeline system was shut off through the vacuum breaker in the conveying pipeline. This vacuum breaker is a motorized butterfly valve that allows air to enter at this point, stopping the conveying of clay from the car unloading twin-nozzle unit.

The complete weighing installation includes upper and lower scale hoppers. The upper scale hopper is necessary to maintain a sufficient quantity of clay to the main scale hopper for weighing accuracy. It is equipped with a high-level cutoff, which automatically opens the vacuum breaker in the conveying pipeline to stop conveying if clay is unloaded and conveyed faster than the equipment following the upper scale hopper can take it away. A screw feeder feeds the main scale hopper from this upper scale hopper. This

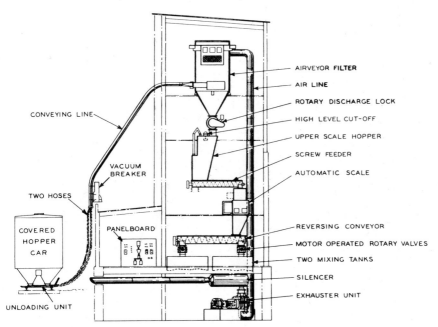

FIGURE 5.24.   Clay unloading to slurry storage. (Courtesy *Tappi.*)

screw feeder fills the main scale hopper quietly and evenly. It should be large and operate slowly so as not to aerate the clay further than the aeration that occurs in the pneumatic conveying system.

As in conveying and storing, special precautions must be taken in the weighing equipment. The main scale hopper is stainless steel to provide a smooth inner surface for minimizing the frictional resistance of the clay against the wall of the hopper to produce a good discharge. To assist the discharge, the hopper is equipped with vibrators, operating only when the discharge gate is opened. This also makes certain that all the weighed material is effectively discharged. Further precaution is made on the discharge gate of the main scale hopper. This gate is power operated and brush sealed so that no material can escape while the weighing is being made. This type of seal is necessary since the clay is extremely fluid and must be contained within the main scale hopper during weighing. The discharge gate is opened on completion of each 250 lb weighing, with the clay being discharged quickly and completely.

From the main scale hopper, the clay is discharged into the lower scale hopper, which acts as a surge hopper. This lower scale hopper should be sufficiently sized to accommodate the total 250-lb dump of the scale. The reversing screw conveyor, which distributes the clay to the mixing tank from the lower scale hopper, is sized to convey at 30–35 tons/hr. If insufficient space is allowed in the lower scale hopper between the inlet to the screw conveyor and the scale discharge gate in its open position to accommodate the 250-lb surge of clay, scale malfunction will be the result.

The mixing of the slurry is accomplished by a propeller-type central mixing shaft combined with baffles built into the side of the mixing tank. It takes approximately $\frac{1}{2}$ hr to complete the mixing of a full batch of slurry to 70% solids. The slurry, after mixing, is automatically drawn from the bottom of the mixing tanks and pumped a distance of about 100 ft to the slurry storage tanks. This entire operation, allowing time for each phase of the process from dry clay in the cars to mixed slurry in storage takes no more than 4 hr and requires only one worker.

The unloading of a whole carload of clay into a single mixing tank is shown in Figure 5.25. This installation, consisting of two separate systems, unloads coating clay to slurry mixing tanks and filler (loading) clay to flat storage. A straight vacuum system is used to unload the coating clay from the covered hopper cars at the rate of 25–30 tons/hr. The filler clay is unloaded from cars through a straight vacuum system to a transfer station, where a low-pressure system blows the clay 400 ft to the flat storage at a rate of 20 tons/hr.

The transfer station, part of the total combination vacuum–pressure system uses two rotary feeders as described in Section 2.7. Since the vacuum leg was designed to operate at 9 in. Hg through an 8-in. conveying pipeline and the pressure leg at 10 psig through a 5-in. conveying pipeline, the differential across the feeders totaled 14.5 psig. Removal of tramp iron was felt

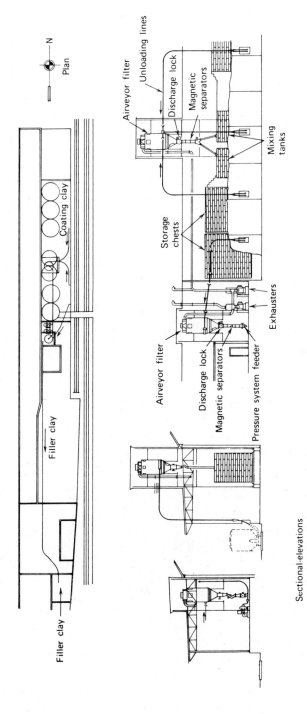

Sectional-elevations

FIGURE 5.25. Clay unloading to dry and slurry storage. (Courtesy Fuller Company, Bethlehem, Pennsylvania.)

to be necessary so the two feeders were separated by a vertical offset spout in which permanent magnetic plate type separators were attached. A bin level indicator was also installed in the spout to shut down the unloading system in the event that more material was being unloaded than was being conveyed away by the pressure system.

The surge in the handling of starch in bulk by the paper mills, like the introduction of the clay systems, has been expanded by the suppliers. The suppliers, influenced by keen competition, undertook the furnishing of bulk handling and storage systems to the mills under a payoff arrangement. The mills pay bagged prices for their starch but receive it in bulk. When sufficient savings have accumulated and the equipment amortized, it then becomes the property of the mill. This arrangement requires the mill to sign a long-term agreement with the supplier, which means that the supplier has a steady customer for his product.

Early systems, having low conveying rates, utilized a portable, twin-feeder, low-pressure type system to convey the starch from the discharge of Airslide cars to storage bins. Later installations used either straight vacuum unloading systems, or a combination of vacuum and low-pressure using a transfer station at grade.

A straight vacuum system with the filter-receiver above the storage bin is shown in Figure 5.26. This 5-in. system, actuated by a 40-hp exhauster unit, unloads the pulverized starch at the rate of 12 tons/hr. To convey the starch from the storage bin to the starch cookers, a separate low-pressure system is used to convey at 6 tons/hr. In the bottom cone of the storage bin, open-top air-activated gravity conveyors are used to aerate the starch to accelerate flow to a twin-screw feeder. This twin-screw feeder discharges into a small surge hopper, which in turn discharges into a dual-air rotary feeder for charging the starch into the conveying pipeline. The system terminates in a weigh hopper with the conveying air dissipating through the cloth filter tubes extending above the roof in a small penthouse. Delivery to the one cooker is direct by gravity spout from the weigh hopper, and to the other cooker, by a horizontal screw conveyor.

The combination vacuum–pressure system is shown in Figure 5.27. In this installation, starch is unloaded from the cars to a transfer point, where a pressure system below the filter-receiver delivers it to the storage bins. To minimize equipment on top of the storage bins, the air required for conveying in the vacuum system is drawn off the top of the storage bins, so that this air dedusts the bins, recirculating through the whole system, obviating the need for dust collectors above the bins.

When conveying calcium carbonate in the same system that handles clay and starch, the carbonate, being a material with variable physical characteristics, will be conveyed at approximately 60–75% of the rate for clay.

In handling English China clay, some difficulty may occur in a pneumatic conveyor. This clay is quite lumpy, and when unloading from boxcars, the large lumps will cause problems at the pickup of the nozzle of the hose and

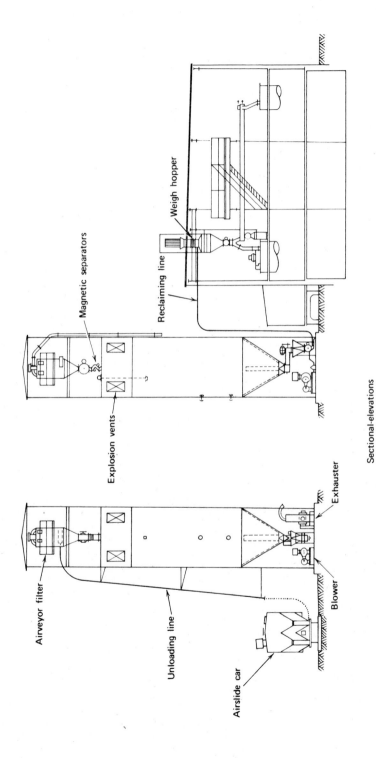

Magnetic separators

Explosion vents

Reclaiming line

Weigh hopper

Airveyor filter

Unloading line

Airslide car

Blower

Exhauster

Sectional-elevations

FIGURE 5.26. Separate vacuum unloading and low-pressure starch reclaiming systems. (Courtesy Fuller Company, Bethlehem, Pennsylvania.)

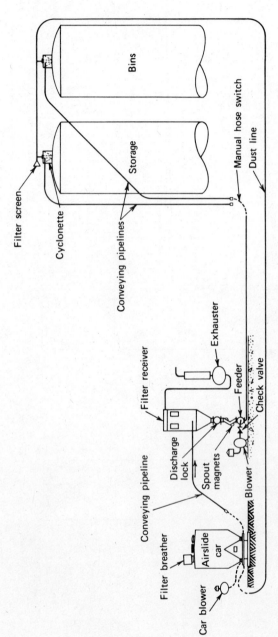

FIGURE 5.27. Combination vacuum and low-pressure starch unloading system.

pipeline system. The moisture content of this clay is much higher than that of domestic clays. The velocity should be higher than normal to avoid buildup in the conveying pipeline bends. Unloading rates will be approximately 60% of those for domestic clays under comparable power and air requirements.

For conveying broke from the paper machine (reel and rewinder end) and trim from cutters and sheeters in the finishing room, a fan- or centrifugal-blower-actuated pneumatic conveyors are frequently used. Two types are available. One uses an eductor or ejector to provide a negative pressure at the intake to the system with a positive pressure discharge. With either the eductor or ejector, no material passes through the fan or centrifugal blower. The other type of system passes the material through the fan. Normally, with this type of system, a cutter is used to cut the broke or trim into pieces small enough so as not to foul the fan blades. The angular blades of such fans are heavily constructed. Without the cutter, the fan is equipped with heavy-duty detachable vanes that perform a chopping operation during conveying. Design of these systems should be left to the equipment manufacturer.

5.8   FEED INDUSTRY

The feed mill receives whole grains, including corn, sorghum, oats, barley, and wheat, and many meals that are by-products of the vegetable oil, grain, starch, and edible food processing industries. The feed mill processes these materials into formula feeds for such livestock as poultry, dairy cows, and dogs. Effective and expeditious bulk handling is important to this industry. On a gross profit margin of 3–8% of the total cost, manufacturing costs amount to only 6–10%. It is therefore imperative to control very closely the cost of materials handling, which is a good part of the manufacturing cost.

Mills, both large and small, use pneumatic conveyors for their unloading operations, as well as in-plant transfer, including finished pellets. The economic use of the pneumatic conveyor is determined primarily by the material to be handled. In many mills both mechanical and pneumatic systems are used in the unloading operation since they both have advantages and limitations that are related to the material being unloaded.

One important advantage of the vacuum unloading system over the mechanical system using a power shovel is the great reduction in handling loss, or shrinkage. In the mechanical unloading of bulk feed ingredients, 1–2% is lost by spillage and by blowing away. In a carload containing 80,000 lb, the loss of material is between 800 and 1600 lb. Whether the mill is large or small, this loss can be substantial since ingredient cost is estimated to be 65–75% of the total cost of a commercial formula feed.

Another important advantage of the vacuum system, and one that receives the most favorable comments from feed mill operators, is its safety.

It entirely eliminates the inherent accident hazard of the power shovel, which is one of the biggest causes of lost-time accidents.

Small mills report savings of from $7.00 to $20.00/ton in their material and production costs by substituting bulk handling for bagged handling. Pneumatic handling from cars and trucks to storage and from storage to processing makes this possible. Among the materials handled are soybean meal, gluten feed, wheat bran, wheat middlings, distillers dried grains, brewers dried grains, and shelled corn, oats, and barley.

Larger mills having a production of 75,000 tons/year and above use both mechanical and pneumatic unloading. Material characteristics alone are not the total guide to the type of system to use. Frequently the conveyance in which the material is received determines the type. Covered hopper cars are slowly but steadily replacing the boxcar in the transportation of bulk grain. Many railroads allow preferential freight rates if 10 or more cars are transported as a unit. When these cars are received in the mill, they must be unloaded quickly and returned to the carrier to avoid costly demurrage charges. Under this operating condition, the dischage of the grain into a pit having a horizontal screw conveyor delivering to a bucket elevator for lofting to the adjacent silos appears to be the most economical and accepted method.

There have been pneumatic unloading systems installed having conveying rates of 3000 bushels of hard grain per hour. They were installed, however, when only boxcars were available. This does not mean that their use for such rates is obsolete. In addition to elevating, they can also convey horizontally through the same pipeline system. If the storage silos are not adjacent to the car-unloading track, then high-volume pneumatic systems have their greatest advantage. With a stationary nozzle connected to a vacuum system in the pit, gravity discharge of the covered hopper cars can be accomplished, together with high-rate conveying.

Most pneumatic unloading systems in the feed industry are of the vacuum type. This allows the unloading of any conveyance, whether it be a boxcar, covered hopper car, or truck. Typical conveying rates for such systems indicate the variations due to differences in materials. For a system employing a 6-in. conveying pipeline, 150 ft long; a 5-in. unloading hose and nozzle; and a 50 hp exhauster unit, the conveying rates would be about as follows:

Oats	1000 bu/hr
Barley	750 bu/hr
Corn or wheat	650 bu/hr
Soft feeds	10 to 13 tons/hr

For a system employing an 8-in. conveying pipeline, 60 ft long; an 8-in. exterior and 7-in. interior telescopic riser, 30 ft long; 7-in. car unloading hose and nozzle; and a 100-hp exhauster unit, the conveying rates would be about as follows:

Oats	2200 bu/hr
Barley	1700 bu/hr
Corn or wheat	1500 bu/hr
Soft feeds	22 to 28 tons/hr

The preceding system will have an operating vacuum of 9 in. Hg. It is desirable to operate at a higher vacuum to attain a higher conveying rate. By increasing the vacuum on this same system to 11 in. Hg, and by utilizing the full potential of the 125-hp motor now needed to drive the exhauster, the increased conveying rates would be about as follows:

Oats	2800 to 3000 bu/hr
Barley	2000 to 2500 bu/hr
Corn or wheat	1800 to 2000 bu/hr
Soft feeds	25 to 30 tons/hr

By comparing the conveying rates of the foregoing two systems, we might wonder why, with all equipment being equal in size except for the motor driving the exhauster, the smaller capacity system is considered at all. The answer lies not in the handling of the whole grain but in the handling of soft feeds such as bran. When feeds having a bulk density of less than 20 lb/cu ft are conveyed, their flow from the bottom cone of the filter-receiver into the rotary discharge lock is impeded by the higher vacuum.

The rotary discharge lock at its discharge is under atmospheric conditions of 14.7 psia. At its inlet, the conditions are 10.2 psia at 9 in. Hg vacuum, and 9.2 psia at 11 in. Hg vacuum. As the rotor rotates within its housing, each pocket as it reaches the inlet contains air at 14.7 psia, which is immediately rarefied, causing a burst of air upward into the bottom cone of the filter-receiver. This countercurrent airflow reacts against the material trying to flow out of the filter-receiver, causing it to hang up. The greater the air differential across the discharge lock, the greater is the possibility of material hanging up.

In the unloading of soft feed ingredients from boxcars by a vacuum system, three factors affect their conveyability: moisture content, oil content, and particle size.

All soft feed ingredients are hygroscopic, especially the meals. Even during transit they will absorb moisture. In the summer months of July and August, this characteristic is much more noticeable than at other periods of the year. Supporting this is the experience of a mill using an 8-in. vacuum system with a 7-in. hose and nozzle in the boxcar. During the winter months, they employ only one man on the system, whereas in the summer months two men are required to maintain good conveying rates, With the higher moisture content, the materials are not so free-flowing and require more manual labor to break down the pile for pickup by the system nozzle. Under ideal conditions, having a rather free-flowing material, one man requires only ½ to ¾ hr more than two men when unloading a 50-ton car. The saving in labor is 1.5 manhours/car. Oil content will have similar results. The higher the oil content, the lower the conveyability. Particle size, extra fine, fine, or coarse, affects both flowability and conveyability. Coarse material will be good, but each decrease in particle size will reduce conveyability. This combination of oil content and particle size is evidenced by three different cars of gluten meal. Normally, this material is a very good one to handle, but startling results were produced. One car containing 78,400 lb was unloaded in 1 hr; the second car, containing 80,000 lb, required 2½ hr; and the third car, also containing 80,000 lb, required 7 hr. This last car contained a very damp meal that required every pound to be broken down for pickup by the conveying system. While conveying rates can basically be determined, their variations from different material characteristics must be expected.

When unloading boxcars of material, the hose and nozzle should be dropped into the car above the bulkhead placed across the door opening that retains the material in the car when the door is opened. The material behind the bulkhead should be unloaded first to avoid spillage. As the level of material is lowered to the floor, the bulkhead can be removed. Then, laying the nozzle on the floor (Figure 5.28) and directing it into the toe of the pile, the pile will be undercut. If the material is free-flowing, it will flow toward the nozzle, maintaining a steady unloading rate. If it is not free-flowing, a wide-bladed (plasterers) hoe is used to break down the pile to the nozzle. With the nozzle submerged, quite a lot of material can be unloaded before the nozzle need be moved. As the material is unloaded, the operator can sweep the floor clean toward the nozzle, and with occasional use of a wide scoop (grain-type) shovel, the car can be unloaded with a minimum of manual labor.

To obtain satisfactory unloading rates, the operator must be a conscientious worker and maintain good work habits. If he sluffs off, total unloading time can increase, sometimes beyond belief. A good example was a mill that continually complained that its vacuum-type unloading system was undersized and incapable of unloading at the designed rate. Each time the vendor's service engineer checked the equipment and operated the hose and nozzle, superior conveying rates were achieved. After many complaints by the mill owner threatening all sorts of mayhem, the service engineer returned—this time, unannounced. On his arrival, he proceeded directly to the unloading site. On the floor of the car lay the hose and nozzle conveying nothing but air (no material). The operator was nowhere in sight. After waiting 15 min, the service engineer made his way to the owner, immediately inviting him to the unloading site. On arrival, the hose and

FIGURE 5.28.   Boxcar unloading of soft feeds. (Author's collection.)

nozzle was still conveying nothing. A half hour later, the operator returned to resume unloading. Needless to say, he was fired on the spot. The new operator, maintaining good work habits, caused no more complaints.

We should not overlook the capability of the low-pressure system in conveying from cars. With greater emphasis on increasing the flowability of materials, many materials that were formerly transported in boxcars can only now be transported in covered hopper cars. Increased flowability allows the material to discharge by gravity and added vibration from the covered hopper cars into mechanical feeders, which in turn discharge into the rotary feeder of the low-pressure system. This arrangement is especially important when high conveying rates are mandatory and/or a multiplicity of discharge points is to be served. With such systems, adequate dust retention devices must be used, not only at the delivery points but also at the car unloading station, where shrinkage or material loss must be controlled.

In-plant handling of feed ingredients is done by both the vacuum and low-pressure systems. After the ingredients are processed, some of the finished products are pelletized for ease of further handling and elimination of dust losses at the points of use. Many producers and processors view the pneumatic conveying of pelleted feedstuffs with suspicion. There are others, however, who favor pneumatic conveying. The average hard pellet can be conveyed with relatively little breakage and attrition. Pellets having a high fat content, however, are very friable and difficult to convey in any system, pneumatic or mechanical.

Pellet breakage can be attributed to three major contributing factors: product formulation, conveying velocity, and free fall in bins and hoppers. Formulation is a factor often overlooked. The ingredients, their preparation, and their treatment determine the hardness and toughness of the pellets. The processor, to take advantage of and use pneumatic conveyors for pellet transferring, is responsible for making a pellet that can be conveyed.

The pneumatic conveying engineer is responsible for maintaining the pellet with minimum breakage and attrition. To do this, the system, either vacuum or low-pressure, must be designed to have velocities in the pipeline in the range of 4000–5000 ft/min. It is equally important to maintain the greatest material load possible in the conveying pipeline. This reduces random movement and impact between pellets, reducing breakage, and stabilizes the conveying velocity. The pipeline should be designed with minimum curvature and to have pipe joints that are smooth and tightly butted.

Approaching the terminal point of the pipeline system, whether it is a filter-receiver in a vacuum system or a cyclone receiver or bin in a low-pressure system, the pipeline should be enlarged to reduce the terminal velocity. This enlargement is shaped like the frustrum of a cone or pyramid. The point at which enlargement begins is just far enough awayy from the terminal point so that the pellets have just enough residual velocity to clear the end of the pipeline.

At the terminal end of the system and any subsequent handling into bins, the amount of free fall of the pellets should be kept to a minimum. Free fall within the bin should be retarded with a spiral or zig-zag chute so that the pellets will flow gently to the level of the material. This reduces pellet breakage as well as stratification within the bin.

5.9   WATER PURIFICATION

Water purification or treatment plants, both public and private, use pneumatic conveyors for such bulk chemicals as alum, activated carbon, soda ash, lime, and ferrous and ferric sulfate. Other chemicals are used in the treatment process, but the amount of usage is insufficient to merit bulk handling.

Beside the basic need for which they are constructed, many public water treatment plants are designed more for esthetic beauty than for ease and simplicity in their bulk material handling systems. Many times, bulk storage bins are spread out, small in volume, and requiring a large floor or storage area. To unload the chemicals and deliver them to this type of an arrangement requires a vacuum system for unloading from the cars or trucks to a transfer point, and then a low-pressure or a combination vacuum and low-pressure system for delivery to the bins. With air delivery to the bins, a dust-collecting system must be added to eliminate dusting at the bins. All of this equipment within the main part of the building requires considerable floor space for its installation and maintenance. It is a complex system, which increases not only the cost of construction of the entire plant, but also the cost of operation and maintenance.

The use of larger storage bins, reducing their number, will allow a simpler handling system having much less equipment. The straight vacuum system, unloading the chemicals from either cars or trucks and delivering the materials to the top of the storage bins, is by far the simplest and most effective arrangement. To deliver to a number of bins, horizontal distributing screw conveyors can be used to good advantage, eliminating the need for costly dust control equipment. Although this arrangement requires a penthouse to house the filter-receiver, this will add only a few dollars to the cost of construction of the building, whereas spreading out the building to maintain a large number of small storage bins will increase the overall cost. Installing an outdoor filter-receiver on the roof can eliminate the need for a penthouse, further reducing overall costs.

In the purchase of pneumatic conveying systems for public works water treatment plants, the pneumatic conveyor usually is included in the general contract. While equipment is quoted to general contractors according to prepared specifications, the contractor in the normal course of preparing the bid will invariably use the lowest price quoted for the system. This means that the best equipment may not be purchased, since the best is not always the cheapest. Considering the many variables involved in the design and operation of successful pneumatic conveying systems, it is recommended that these systems be purchased under separate contract. Then various bids can be tabulated according to equipment proposed as well as price, not by price alone. Then and only then can the right purchase be made. Many years of experience have indicated that this industry, especially the public works segment, needs the advice of the material handling engineer not only to reduce costs, both original and operating, but to simplify the handling operation.

The three most used chemicals, alum, lime, and soda ash, can be handled very expeditiously in a single system. Pneumatic conveyors, through their ability to be self-cleaning, minimize contamination between them. There are, however, several points in the systems where special considerations must be made. In the handling of alum, rubber hose bends should be used in the conveying pipeline since this material tends to build up and smear in long-radius steel bends. Below the filter-receiver, gate locks are recommended over the rotary feeder discharge lock. The alum will build up on the interior of a rotary discharge lock, causing a smearing and binding action that will result in malfunction. While rotary discharge locks have been used, the clearances between the rotor and the body had to be widened to eliminate the smearing action from the buildup. When this was done, air leakage through the lock increased perceptibly. This additional air leakage and blowback of air into the bottom cone of the filter-receiver decreased the life of the filter media, increasing maintenance. It also affected system operation due to this air loss. In systems handling pebble lime, the pipe bends when fabricated from steel pipe must be reinforced against wear from the abrasive action of the lime.

The other three bulk chemicals, activated carbon and ferric and ferrous sulfate, require other special considerations. Activated carbon, which has a bulk density of only 15 lb/cu ft, is best handled in a vacuum system. It is a most difficult material to contain, so by the use of a vacuum system, all leakages are in, not out. Carbon dust can cause a most difficult housekeeping chore. The filter media should have a ratio of not greater than 1:1 for woven media, or 2:1 for felted. Operating vacuum should be limited to 9 in. Hg. This will minimize blowback through the discharge lock into the bottom cone of the filter-receiver.

Ferrous sulfate should be handled in a completely separate system. Any mixture of this material with hydrated lime, or any type of lime, will cause a reaction that will blind the filter media in the filter-receiver and will cause operational difficulties in the discharge lock. Ferrous sulfate is very hygroscopic and is heat sensitive. When conveying this material, the temperature of the conveying air should not exceed 125°F. Vacuum conveying is successful, since a vacuum system tends to reduce moisture in material rather than add moisture, as a pressure system does. With this criterion, it is best to avoid using pressure systems of any type for this material. For storing in bulk, a dry atmosphere is required.

Ferric sulfate, while not as difficult to handle and store as ferrous sulfate, should receive the same considerations. It, too, is hygroscopic, but it may be conveyed in the same system with other materials. Caution, however, must be used.

For conveying grit and sludge cake in water and wastewater treatment plants, the pneumatic ejector (blow tank) is most applicable. Two types are used: the top loading for handling semidried solids and sludges and the bottom loading for handling wet, free-flowing sludges and slurries.

5.10 POWER INDUSTRY

With the advent of the energy crisis in the 1970s and the industry's changeover from oil to coal as fuel, pneumatic conveyors have received wide acceptance in the handling of fly ash, pebble lime, and pulverized limestone. The handling of pulverized coal has been a well-accepted fact since the early 1920s. Although not entirely applicable to the power industry, it is worthy of note that in the mid-1920s, two companies, Fuller Engineering Company and Combustion Engineering Corporation, developed pulverized fuel equipment for firing steam locomotives. They both used a turbo-blower set that supplied the air for conveying the fuel from the tender to the locomotive's firebox.

Although fly ash has been conveyed intermittently over the past 50 years, the changeover to coal has resulted in collosal baghouses and precipitators for air pollution control that require the continuous withdrawal of the ash from their collection hoppers. In any baghouse or precipitator, the fly ash collected is deposited in sometimes a few and sometimes many hoppers. The ash is allowed to collect in these hoppers. Periodically, they must be emptied.

With fly ash being an extremely abrasive material, abrasion-resisting equipment had to be developed to give long life expectancy of equipment together with dependable, nearly continuous operation. The pneumatic conveying industry was equal to the task. Systems used for fly ash conveying include vacuum, low- and medium-pressure blow tanks, fluid solids pumps and air-activated gravity conveyors.

The vacuum system uses specially designed intake valves that have no close-fitting moving or rotating parts. (Rotary feeders for such service should never be used; they would not stand up under such punishing service and maintenance would be astronomical.) The intake valves are automatically sequenced so that a hopper is emptied into the conveying pipeline while all others are closed. Through adequate sequencing, a normal conveying capacity is attained, which keeps equipment sizes within reason. To accelerate flow out of the hoppers, the vacuum system should have a preload arrangement whereby a vacuum of 3 in. Hg is maintained within the pipeline system.

Vacuum systems are used mostly where headroom between the collecting hopper outlet and the floor is insufficient to accommodate the blow tank or feeder arrangement. They are kept to minimum length and usually terminate in a transfer station for delivery of the ash to storage or a loading-out silo. The transfer station includes the usual filter-receiver for the vacuum system. The ash on discharging from the filter-receiver is deposited directly into a blow tank, with the blow tank not only activating the pressure system to silos, but also acting as the airlock for the vacuum system. Since the emptying of the blow tank into the pressure conveying pipeline is an intermittent operation, the bottom cone of the filter-receiver must also act as a holding hopper. It should be designed so that sufficient volume is afforded to accommodate the accumulated incoming material while the blow tank is discharging. Also, care should be taken in designing the size of the dust collector atop the silo to accommodate the bubbles of air from the blow-tank system.

Pressure systems use either small blow tanks or feeders such as the Perma/

lok III airlock. They too are sequenced so that only one blow tank or feeder is feeding material into the conveying pipeline. One thing to remember, however, is that the amount of material collected in the hopper should never be greater than the volumetric capacity of the blow tank or feeder. In fact, it would be wise to set up the sequencing so that the amount of ash collected in the hopper is no greater than 75% of the volumetric capacity of the blow tank or feeder. By adhering to this recommendation, malfunction from insufficient sequencing is reduced to zero.

For the conveying pipeline, ductile iron should be used. Its abrasion-resistant quality will give long life with minimum maintenance. To avoid pipeline distortion due to expansion and contraction due to temperature variations between hoppers, expansion joints should be installed in the piping between the hoppers.

Where quantities of ash are sufficient to maintain a continuous flow of material, fluid solids pumps are applicable to convey the ash to the storage or loading out silos. Air-activated conveyors collect the ash from the baghouse or precipitator hoppers to a central hopper above the pump. With this arrangement, however, headroom could be a problem.

For obvious reasons, all systems whether vacuum or pressure, should have stand-by air supply systems. Powerhouse operation is not an intermittent operation. So long as the boiler is making steam, fly ash collection must be maintained.

For flue gas desulfurization, pneumatic conveyors adequately handle the required pebble lime and pulverized limestone. To unload these materials from covered hopper cars, vacuum, low- and medium-pressure blow tanks and fluid solids pumps are used, depending on material characteristics. In-plant systems include the blow tanks for both materials and the fluid solids pump for only the pulverized limestone.

Much research is under way in the development of the fluid-bed boiler. The main criteria for its success is a reliable and accurate coal-feeding installation. The coal, mixed with limestone, must be spread evenly across the entire bed area. This area in large installations will require many feed points for combustion efficiency.

From the experience attained in feeding coal to kiln burners and flash calciners in cement mills (Figure 5.2), the latest design of fluid solids pump together with steam splitters is being used in a 20-MW pilot plant. This method of fuel injection was chosen because of the pump's positive displacement, pressure seal, and variable-feed capability. One design proposal for a 150-MG installation envisions 240 feed lines activated by 10 pumps and 24-way stream splitters.

5.11 FERROUS METALS INDUSTRY

In the past 20 years, iron and steel foundries have benefited greatly in the handling of their raw materials and in in-plant handling of shake-out sand to reclaimers and raw materials to their mullers. The blow tank, both low- and medium-pressure, utilizing low conveying velocities in the pipeline system, has made this possible.

Raw bonding materials such as bentonite, cereal binders, and seacoal were among the first materials to be handled pneumatically. They were received in bulk with the railcars unloaded by vacuum-type systems. More recent developments include pneumatic injection of materials including coke breeze, nickel oxide, burnt lime, and dolomitic limestone through tuyeres into cupolas, and blast and decarburization furnaces.

The application of pneumatic conveyors to this industry, including design, conveying rates, operating vacuums and pressures, pipeline velocities and systems, and automation should be put in the hands of the experienced. All materials are abrasive, and in development all systems require close cooperation between the metallurgist, operating personnel, and the pneumatic conveying engineer to arrive at the best workable arrangement. Systems mostly used are blow tank and low-pressure using abrasion-resistant airlocks and feeders. Operating pressure and air volume information is predominantly proprietary.

Through the use of pneumatic conveyors, material losses are kept to a minimum; high accuracy through automatic weighing, either batch or continuous, is attained; and material flow is adequately monitored. These operating characteristics contribute greatly to economic savings and uniformity of product.

In the manufacture of coke for the industry, coal is heated to about 1400°C without allowing air to burn it, expelling volatile matter including ammonia. The ammonia is processed into ammonium sulfate, used as a fertilizer and for fireproofing. The final stage of its manufacture is centrifuging to reduce its moisture content to about 2.0%. The moisture content for storing and for future use should be below 0.5%. A low-pressure system using heated air adequately performs the dual purpose of drying and conveying. The system should be sized to use minimum air requirements and low conveying velocity to maintain its crystalline structure, so necessary for fertilizer, to avoid dust losses.

5.12   TRANSPORATION INDUSTRIES

Automotive Industries

The application of pneumatic conveyors for truck transportation dates back to the 1920s. Among the first applications was the transportation of malt from the team tracks of the New York Central and New York, New Haven & Hartford Railroad's yard at Mott Haven, the south Bronx, New York, to the upper Manhattan brewery of the Jacob Ruppert Brewing Company. The trucks were Mack Bulldogs* having four cylinder engines, solid rubber tires,

---

* Registered trademark, Mack Trucks, Inc., Allentown, Pennsylvania.

and chain drives from the rear intermediate axle to the wheels. Carrying about 300 bu per load, they used a vacuum system to transfer the malt from the boxcars to the tank, and delivered to a bucket elevator at the brewery.

In the mid-1940s, trailers were developed to load 500 bu into their tanks with on-board exhausters driven by a gasoline engine, dust filters, and ribbon screw conveyors to feed malt out of the rear discharge chute. These units were quite high, but still able to pass under all public overpasses, except, a private one, the ornamental ironwork at the gate to the brewery. The truck, in its initial loading, took on about 550 bu of malt in 10 min at a Central Railroad of New Jersey team track in lower Newark, New Jersey. The ride to the brewery was uneventful until it started to back into the brewery yard. The iron grillwork that spanned the yard entrance and proudly displayed the brewer's name was about 18 in. lower than the top of the trailer. The bucket elevator that was to receive the malt was about 40 ft away. It is a shame film and camera were unavailable to record the ensuing events.

When the predicament became known, many people, including the brewery's president, congregated to offer all kind of assistance and advice. It would seem logical to rent a portable belt conveyor (similar to those used in coal yards) to transport the malt from the truck discharge to the elevator boot. The "brass hat" would have none of that. Instead, he sent several men to the boiler house to get a number of coal chutes. These were placed from the truck to the elevator, supported by wooden beer cases. Then he spaced men at 6-ft intervals, alternately on each side of the chutes, each equipped with a heavy-duty corn broom. As the malt discharged from the truck, it was swept along by the men to its destination—even eighteenth-century material handling was superior to this. All 2800 bu from the car were broomed into the brewery. Moral: Check *all* overpasses for road clearance.

The early entrepreneurs were always seeking out applications for their product. Since coal was then used as the fuel for tenement heating, an ash removal truck was proposed (Figure 5.29) to pick up the ashes with a vacuum

FIGURE 5.29. Ash collection truck circa 1929. (Author's collection.)

FIGURE 5.30.   Ash collection truck circa 1942. (Author's collection: he faces the camera.)

system and deposit them in a tank for disposal. Separate trucks were to be used, one for the ashes, the other for the power plant (gasoline engine and exhauster) and a dust retention filter that consisted of multiple flat muslin-enveloped screens. To activate the ash truck, a flexible hose connected the vacuum source to the tank. The ash tank was raised through a mechanical rig for discharge through a rear opening. The rig operated through a power takeoff from the truck's transmission.

Thirteen years later, an ash removal truck (Figure 5.30) was built for an inner-city housing authority to remove ashes from the boiler rooms of housing projects. The truck was a Mack model EQU (cab forward) chassis equipped with two power takeoff attachments to the transmission: one to operate a twin-cylinder hydraulic hoist for raising the tank for dumping, and the other for driving the exhauster (blower) through a splined driveshaft with universal joints at both ends and multiple v-belt drive. The unit gave good service, but should have been equipped with a safety interlock to prevent the hydraulic hoist from operating when the rear discharge door was in a closed position. Although the operating instructions explicity called for the discharge door to be opened before hoisting the tank, the operator one day failed to comply. When the tank was raised, the load all shifted to the rear, causing the entire truck to nearly stand on its tail; the chassis at about 40° from the horizontal. Hooking another truck to the front solved the problem. Moral: Prevent human errors.

The application of pneumatic conveyors to bulk feed trucks began in the early 1940s in California. A feed manufacturer built the first and later an entire fleet. The main reason was that during World War II bags were in very short supply. Economic advantages were readily acknowledged, causing the rapid deployment of bulk feed delivery to farmers and feedlot operators in the early 1950s. In 1952, Sprout, Waldron & Company, Muncy, Pennsylvania, developed a pneumatic truck and the prototype was put into service at the Central Jersery Farmer's Cooperative Association, Hightstown, New Jersey. After several modifications, it was declared operational in 1953. Sprout, Waldron then established an assembly line for production. Two types were produced; low-pressure, using a rotary feeder to feed the conveying pipeline, and pressure-differential (blow tank), to operate below 15 psig.

Also during this period, several industries, notably sugar refining, had special trailers built for pneumatic bulk sugar delivery. The design was based on the blow tank principle. Three 10,000-lb-capacity tanks of stainless steel construction were activated by an on-board blower driven by a gasoline engine. Included was a dust collector to collect the dust from the conveying operation. Its purpose was to reduce any caking or "hang-up" tendency in the storage bins. Six-inch sanitary hoses connected the discharge pipe to the conveying pipeline and the dust return line from the storage bins to the dust collector inlet pipe. Unloading rates were 20 tons/hr over a conveying distance of 150 ft.

The development of the Airslide pump and its application to the transportation of well driller's mud for the oil well–cementing industry in the early 1950s proved the feasibility of pneumatic unloading for over-the-road highway equipment. Transported on flatbed trucks, the pumps could deliver the mud to where it was actually required. They were mounted on a slope of 8–10°, so that the material would flow to one end, and discharged through the pipeline by the blow-tank principle, also below 15 psig. Refining the principle to increase capacity by joining two pumps together to form a single tank with a centrally located common outlet, an over-the-road self-unloading trailer developed. However, it could handle only fluidizable material.

To handle coarse as well as fluidizable material, the pressure-differential smooth bore tank trailer was developed and refined (Figure 5.31). Materials transported and pneumatically unloaded include cement, sand, clay, fertilizer, salt, sugar, flour, plastic pellets, and dispersion resins. Construction materials include aluminum and steel. Today aluminum seems to be the preferred material because of its lighter weight, which affords higher payloads or reduced overall weight of the vehicle. Discharge lines are usually 4 in. Most have blowers, driven by either gasoline engine or an electric motor mounted integrally with the trailer. If a single destination is used with multiple tank trailers, an on-site blower unit is more economical.

The "cementers" as the trailers used for hauling cement are called, use air pressure at 14.5 psig and an air delivery of 400 cfm. Other materials may call for variations in air pressure and delivery, but the pressure should never be greater than 14.5 psig.

FIGURE 5.31.   Pressure-differential self-unloading truck. (Courtesy Penske Tank Company, Minneapolis, Minnesota.)

The self-unloading tank trailer is advantageous to both large and small users, especially if no rail siding is available. The truck driver is also the trailer operator, unloading it without additional plant labor. Savings are also accrued because there is minimal spillage of material while unloading. The economical traveling distance can approach 250 miles. Beyond this, rail transportation is cheaper.

At the user's plant, easy accessibility to the unloading site should be mandatory. The trailer should be able to park adjacent to the storage facility to which it will discharge its cargo. This will keep the pipeline length to a minimum to allow the highest rate of discharge. Dust collectors atop the bins should be generously sized for adequate dust control. The level of material within the bin or silo should be indicated so that the operator can be sure the bin can accept the load of material. A red and green light at the unloading site, actuated by a material level indicator placed in the bin at a point where a trailer load can be accepted without overfilling, is a simple safety procedure. If the light is red, don't unload; if green, proceed.

Railroad

Since its earliest inception and prior to the 1950s, rail transportation always had the problem of how to get material out of the cars. From the earliest house car or boxcar to the present there has been the demand that one must go in and move the material out, either manually, mechanically, pneumatically, or by rotary or oscillating dumping machinery. When coal became a prime commodity for rail transportation, the open-top hopper with bottom discharge became a reality. In the early 1920s the transition from bag to bulk shipments fostered the design of cars other than box and open hopper to afford weather protection to the lading and gravity discharge. The first industry to benefit from this transition was the cement industry with the development of the covered hopper car. At the time, equipment was available to transport the material from the cars, mechanical and pneumatic (fluid solids pump and blow tank). After World War II the development of the covered hopper was such that today many commodities, including grain, use this mode of transportation.

The need for complete protection and self-discharging prompted the application of the air-activated gravity conveyor to railroad car design. This resulted in the development of General American Transportation Corporation's Airslide car in the early 1950s. Although it was developed for bulk flour transportation, other materials handled include sugar, starch, gypsum, boric acid, some types of clay, and many others. The material on discharge is conveyed by either a vacuum system or a low-pressure system using a portable unit with rotary feeders to feed the pipeline system.

Although covered hoppers have been getting larger each year (up to 5850 cu ft and 100-ton capacity), they all discharge either by gravity out of their three and four hopper outlets or by built-in pneumatic nozzles or outlet gates for vacuum conveying.

In the early 1960s, development of the self-unloading railroad car started to percolate. The success of the over-the-road highway trailer spurred on the car builders. The early proponents included ACF, GATX, North American Car, and Union Tank. Two types of pressure-differential (blow tank) cars eventually developed. ACF Industries developed their low-pressure 5 psig Center Flow car, which was a modification of their standard Center Flow* covered hopper. Materials transported and unloaded include flour, cement, plastic pellets, diatomite, and clay. North American Car, in conjunction with the Transportation Equipment Division of Butler Manufacturing Co. (predecessors of Penske Tank Company), patterned their car after Butler's pressure-differential over-the-road highway trailer to operate below 15 psig. Today this Pd† covered hopper (Figure 5.32) is available in three sizes, 2785, 3915, and 5150 cu ft, all equipped with 100-ton roller bearing trucks. Many materials are shipped in these cars, including some difficult to handle such as caustic soda and titanium dioxide.

In 1981, ACF Industries introduced their latest pressure-differential covered-hopper railcar, the ACF Pressureaide‡, designed for the transport of flour and similar commodities. It operates at no greater than 14.5 psig, and capacity of the car is 5000 cu ft.

Also in 1981, General American Transportation Corporation introduced

---

* Registered trademark, ACF Industries, Inc., St. Charles, Missouri.
† Registered trademark, North American Car Corporation, Chicago, Illinois.
‡ Trademark, ACF Industries, Inc., St. Charles, Missouri.

FIGURE 5.32.   Pressure-differential self-unloading covered hopper car. (Courtesy North American Car Corporation, Chicago, Illinois.)

their Pressure Slide* pressure-unloaded covered hopper car. This car, it appears, is a further development of the Airslide pump to cover not only self-unloading over-the-road highway trailers but also railcar transportation. It uses an air-activated gravity conveyor to fluidize the material for flow to a central outlet, and with its circular cross section, operating pressures up to 30 psig are possible for discharging. This could promote longer conveying distances between rail siding and material destination. Capacity of the car is 3000 cu ft.

Marine

One of the earliest marine applications of pneumatic conveyors was the unloading of grain from ships and barges. First attempts used only a pipe dropped into the hold without the addition of air to accelerate the grain into the conveying pipe. At best, it was not too successful. The development of the air-sleeve suction nozzle in the early 1890's by F. E. Duckham in England solved the problem and gave great impetus to the operation. The nozzle consisted of a sleeve around the conveying pipe that allowed air to be drawn into the material so that the material would be picked up in a state of loose suspension (Figure 5.33). By having the bottom of the sleeve an inch or two below the bottom of the conveying pipe, the flow of air through the sleeve from atmosphere picked up the material for conveying through the pipe. The area between the outside of the conveying pipe nozzle and the inside of the sleeve should be about 1.6 times the net area of the nozzle pipe.

Further development was made in 1906 by A. H. Mitchell, engineer for the London Grain Elevator Company, when he tapered the boom and te-

* Registered trademark, General American Transportation Corporation, Chicago, Illinois.

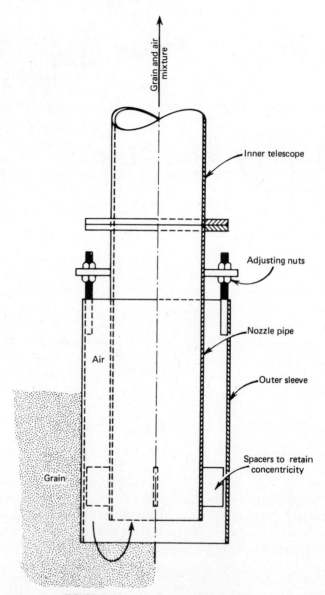

FIGURE 5.33.   Air-sleeve suction nozzle.

lescopic pipes. The nozzle at the intake of the system had the smallest diameter, the inner telescope slightly larger, and the outer telescope and boom pipes still larger. His findings lowered power consumption dramatically from 2.5 to 1.5 hp/ton. It must be noted, however, that power consumption will vary according to the conveying distance involved, i.e., the length of boom and telescopic riser pipes.

Most early units were stationary, either on shore or mounted on barges or specially constructed boats. Around 1910, traveling towers were built to unload the ships by moving the tower rather than warping the ship along the wharf. Delivery of the grain to storage was done by belt conveyors. The axiom, then and now, unload pneumatically, convey mechanically to storage.

The marine tower's pneumatic equipment consists of the exhauster unit, receiving unit with airlock, horizontal boom pipes capable of being raised from a level position to about 80° above the horizontal and swinging through an arc of 180°, a boom bend that is reinforced against wear at the end of the horizontal boom and top of the telescopic riser, a flexible rubber hose between the boom bend and the top of the telescopic riser, and the telescopic riser with the marine-type (air-sleeve) nozzle at its bottom. Boom pipes and telescopic risers are raised and lowered by means of winches or hoists. Some towers include weighing equipment to ascertain correct receipt of grain.

Early towers used cyclone receivers for separation of the grain from the airstream. With pollution control increasing worldwide, substitution of a filter-receiver is becoming the standard.

A marine tower unloading wheat at a designed rate of 3000 bu/hr (90 tons/hr) used a single 12-in. boom and outer telescope pipe, 11-in. inner telescope pipe and nozzle, and four 5-in. flexible metal hoses and nozzles for cleanup. The exhauster was driven by a 210-hp diesel engine through a 2.46:1 ratio gear reduction unit. On the dip (material available to the nozzle at all times), unloading rates were 120 tons/hr with momentary rates as high as 144 tons/hr. Saturation at 90 tons/hr was 1.6 cf air per pound of material and at 120 tons/hr, 1.27. Velocities under full load were 121 ft/sec in the 11-in. nozzle and inner telescope pipe and 100 ft/sec in the 12-in. outer telescope and boom pipes. Velocity in the 5-in. clean-up hoses was 146 ft/sec. Brake horsepower used by the exhauster at 10-in. Hg vacuum was 197, hp/ton 2.19.

A marine tower unloading wheat at a designed rate of 5000 bu/hr (150 tons/hr) and momentary rates of 7500 bu/hr (225 tons/hr) used a single 16-in. boom and outer telescope pipe, 15-in. OD inner telescope pipe with 0.250-in. wall thickness, 14-in. nozzle, and six 5-in. clean-up hoses. Two exhausters, operating in tandem, were each driven by a 150-hp diesel engine through a 2.17:1 ratio gear reduction unit. Saturation at 5000 bu/hr was 1.59 cu ft air per pound of material and at 7500 bu/hr, 1.06. Velocities were 124 ft/sec in the 14-in. nozzle, 115 ft/sec in the 15-in. inner telescope pipe, and 101 ft/sec in the outer telescope and boom pipes. Velocity in the 5-in. clean-up hoses was 162 ft/sec. Brake horsepower used by each exhauster at 9-in. Hg vacuum was 140, hp/ton, 1.86. By increasing the vacuum to 11- or 12-in. Hg and the power to 200 hp for each exhauster, unloading rates of 300 tons/hr are possible. To attain this, the outer sleeve of the nozzle must be raised to about a half inch below the bottom of the inner 14-in. nozzle pipe.

These two towers were put into service about 30 years ago. They served the Victory and Liberty class of ships from World War II. The Victory class were 455 ft long with beams of 62 ft and could carry 170,000 bushels of grain in their five holds. In the mid-1960s oil tankers were a glut in the marine transportation field. Many were cleaned up and used for grain transportation. With no open hatches such as the Victory and Liberty classes, entry into the holds had to be through manholes that did not give free movement to a telescopic riser. This problem led to the design of marine towers using two smaller-sized booms and telescopic pipes, both activated by a single exhauster and receiving station. With the size of the tankers increasing, the necessity of a telescoping boom pipe became necessary to reach the outer manholes. With the size of general cargo ships also increasing, the telescoping booms became necessary to reach their wider beams.

A two-boom marine tower unloading wheat at a designed rate of 150 metric tons/hr (5500 bu/hr) uses two 10-in. boom and outer telescope pipes and 9-in. inner telescope pipes and nozzles, both terminating in an 8-ft, 0-in. two-stage cyclone receiver. A single rotary positive blower (exhauster) was driven by a 350-hp electric motor. On the dip, unloading rates reached 170 metric tons/hr. Saturation at 150 metric tons/hr was 1.29 cu ft air per pound of material, and at 170 tons, 1.08. Velocities under full load were 125 ft/sec in the 9-in. nozzle and inner telescope pipe and 98.5 ft/sec in the 10-in. outer telescope and boom pipes. Velocity in the two 6-in. clean-up hoses was 142 ft/sec. Brake horsepower at 11-in. Hg vacuum was 318, hp/ton, 1.93.

As we analyze the preceding tower, we note that the velocity in the 9-in. nozzles is quite high. Latest thinking in velocity rates is to keep them in the neighborhood of 100–110 ft/sec. The reason for this tower's high velocity was that when only one boom was conveying grain (the other nothing but air), the velocity had to be great enough to put a false load on the one conveying nothing so that the other boom received sufficient air and vacuum to continue conveying material. To alleviate this problem and use lower velocities, GEC Mechanical Handling has developed a sensing system (patent applied for) that automatically partially shuts off the no-conveying boom with valves mounted at the inboard end of the boom pipe.

With the increase in general cargo ship's size and greater cargo space, the need for larger marine towers also increased to keep turnaround time in port to a minimum and within economic reason. A rail-mounted (traveling) ship-discharging tower with one pipe for 600 tons/hr (Figure 5.34) illustrates the length of boom pipes necessary for these larger ships. Boom and telescopic riser pipes range from 24 to 26 in. in diameter. The length of boom pipes on such towers reaches 80 ft and the total height of the telescopic risers, 125 ft. Considering the total conveying distances involved, these towers are still economical using approximately 1.5 hp/ton of grain unloaded.

For higher unloading rates, the bucket elevator marine leg with pneumatic clean-up system is used (Figure 5.35). The marine bucket elevator

FIGURE 5.34.   600-ton marine tower unloading heavy grain. (Courtesy Buhler–Miag GmbH, Braunschweig, Federal Republic of Germany.)

unloads at 1100 tons/hr, picking up the grain in the hold until its operation is impeded by insufficient grain available to it. With the low angle of repose of grain, the unit operates for considerable time before this condition occurs. Then a front-end loader is lowered into the hold to push the grain to the mechanical leg. The pneumatic clean-up units (up to 200 tons/hr each) can operate during the time of mechanical unloading, augmenting the unloading rate, or by cleaning up the grain in the adjacent holds.

Marine towers are used for unloading many other materials, including alumina, soda ash, lime, cement, gypsum, and cereals. Some of these materials are not free flowing. Without some type of exterior assistance, the simple marine nozzle (grain type) will drill holes in the material, which we call "woodpeckering." To alleviate this problem, rotating feeders are attached to the nozzle so that material is loosened and directed to the nozzle through spirals attached to the rotating mechanism. Another method is the use of rotating disks to break down the material from inert to fluid state, and to direct the material to the nozzle intake. Although this is a relatively new development in vacuum conveying, it can be traced back to the first

FIGURE 5.35.   Mechanical marine leg with pneumatic cleanup. (Courtesy GEC Mechanical Handling Limited, Melksham, Wiltshire, United Kingdom.)

portable fluid solids pumps used to unload boxcars of cement in the early 1930s.

The conveying of cement by vacuum had its inception in the early 1960s through research by the Research Institute of the Cement Industry in Dusseldorf, West Germany, and Claudius Peters AG, in Hamburg, also West Germany. The results obtained from this work were most interesting. Vacuums ranged up to 18 in. Hg; pickup velocities ranged as low as 60 ft/sec; and saturation as low as 0.7 cu ft of air per pound of material. The high conveying rates together with the high vacuums indicate that to attain these results, the material must be in a state very close to complete fluidization. In other words, what we have is a solid, through fluidization, acting like a liquid. This research has resulted in marine towers unloading cement by vacuum and delivering to on-shore storage through fluid solids pumps (Figure 5.36).

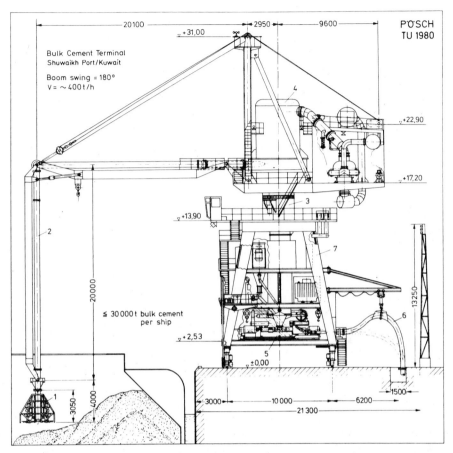

FIGURE 5.36. Marine tower unloading cement by vacuum system. (Reprinted, by permission, Poesch, H. "Pneumatic Unloading Equipment for Bulk Cement Carriers," *Bulk Solids Handling* 2 (1982), No. 2, pp. 301–307.)

The pneumatic equipment for the maximum unloading rate of 400 tons/ hr consists of (1) the suction nozzle equipped with two rotating disks; (2) telescopic riser and boom pipe for vacuum conveying to the filter-receiver; (3) constant-head airlock (patented by Claudius Peters AG), which discharges the cement at atmospheric pressure to the fluid solids pump; (4) vacuum filter-receiver; (5) fluid solids pump; (6) pipeline to silos, including flexible rubber hose attached to equally spaced intake tees to accommodate various positions of the tower when unloading from various holds; and (7) the tower itself, which travels on rails, and the swivel crane that houses the vacuum unloading equipment. What makes these towers practical and successful is that they unload a material whose physical characteristics are universally uniform. Wherever cement is manufactured, the standards of the American Society for Testing and Materials are usually followed, hence product uniformity.

Self-unloading ships and barges have been in use for about 30 years. The material most associated with this type of operation is Portland cement. This operation began as the result of the over-the-highway self-unloading trailer's adaptation to short-haul transportation (below 150 miles). By using water transportation from the cement mill to distribution stations, transportation costs were reduced considerably.

Specially designed ships and barges for cement only use various combinations of the air-activated gravity conveyor, mechanical collecting screw conveyors, and fluid solids pumps. Barges vary in size from 4500 to 90,000 bbls; ships, both ocean and Great Lakes, vary in size from 18,000 to 60,000 bbls.

When a ship or barge is used as a common carrier (cement outgoing, other returning) and on-board unloading equipment is a necessity, a dragline scraper has been used with much success to bring the cement to the fluid solids pump. To permit clamshell unloading or cargo slings for packaged goods on the return trip, the trolleys and cable for the dragline are stowed under the hatch openings. Barge sizes are around 7500 bbls and ship sizes up to 80,000 bbls.

A recent development (1976, H. W. Carlsen AB, Malmo, Sweden) in self-unloading ships is the use of mechanical conveyors (twin screws) to transfer the cement from the hold to suction bins at the ship's sides for vacuum conveying to a combination receiver–blow tank. The screws are traveling units, lowered to trim the top of the cement to the suction bins. Two receiver–blow tanks are employed so that while one is filling, the other is discharging to the shore station. Ship size is 18,000 tons (96,000 bbls). Although these ships are basically cement carriers, they are capable of carrying any type of return cargo. The hold is not self-cleaning, however. Because small residue will remain in the bottom of the hold, a front-end loader is needed for complete cement removal.

In all cases, whether fluid solids pumps of blow tanks, discharge is made through permanent piping to the deck of the barge or ship. Flexible rubber hoses connect this piping to the on-shore pipeline. On-shore conveying distances have reached 1500 ft. Air for the air-activated gravity conveyors is usually supplied by rotary positive-pressure blowers; the fluid solids pumping system by rotary sliding vane-type compressors; and blow tanks by either rotary positive-pressure blowers or air compressors with and without air receivers—all permanently installed in the barge or ship.

For unloading cement and other fluidizable materials from barges and ships with portable equipment, two differing pneumatic approaches are used: the fluid solids pump and the venturi-induced vacuum load and blow tank discharge principle.

In operation, the fluid solids pump is connected to a compressed air supply through a flexible rubber hose and an electric supply cable to supply power to the pump screw motor, the individually powered rubber-tired wheels, and the horizontal rotating disk or disks at the front end of the machine to feed the material into the open end of the pump screw. Its discharge is connected to permanent on-shore piping with another flexible rubber hose. The "wand" in the hand of the operator (Figure 5.37) controls movement of the machine forward, backward, right, or left. Low-voltage electric cable connects the control wand to the relay cabinet, with higher-voltage cables connecting the relay control cabinet to the pump motors. The pump is principally used for conveying distances of up to 500 ft at conveying rates approaching 75 tons/hr.

Although the portable fluid solids pump dates back to the early 1930s, the application of the venturi-induced vacuum load and blow tank discharge to ship and barge unloading is a development of the early 1970s. The Dock-sider* (Figure 5.38) is unloading cement from the hold of a 25,000-ton-capacity ship. This unit features dual transfer vessels, each having a capacity of 50 cu ft, with vacuum pump suction and dual spinners alongside the suction nozzle for material aeration. Conveying rates for single-transfer vessels vary from 50 to 100 tons/hr; the dual-transfer vessels from 150 to 200 tons/hr over conveying distances up to 1500 ft. These rates are based on free-digging time; excluded is time for trimming or warping.

FIGURE 5.37.   Portable fluid solids pump. (Courtesy Fuller Company, Bethlehem, Pennsylvania.)

* Trademark, of the Cyclonaire Corporation, Henderson, Nebraska.

FIGURE 5.38. Docksider barge and ship unloader. (Courtesy Cyclonaire Corporation, Henderson, Nebraska.)

These units are swung aboard the ship or barge by either dock-mounted cranes or ship's gear. The same crane or gear is then used for handling the unit in the ship's hold. In operation, the unit is connected to a compressed air supply through a flexible rubber hose and hydraulic control cables to power the spinners and to operate the automatic control valves. This automatic control system raises and lowers the unit in the material, minimizing handling by the crane operator. Discharge from the unit is through flexible rubber hoses to the permanent piping on shore.

5.13 MISCELLANEOUS APPLICATIONS

Minerals Industry

In years past, the minerals industry had few applications for pneumatic conveyors. With the recent introduction of abrasion-resistant feeders and ductile iron pipe, more materials can now be handled than heretofore. Ore concentrates and tailings conveying is a distinct possibility through the use of medium- and high-pressure blow tanks. The physical characteristics of each material must be studied thoroughly with great emphasis on its abrasiveness.

One process that has enjoyed successful application of pneumatic conveying is taconite processing to convert a sedimentary chert with 20–35%

iron by crushing to powder, concentrated to 65% iron and pelletized for shipment to the steel mills. Vacuum systems, fluid solids pumps, and blow tanks are used in car unloading to storage, storage to mixing plant, and mixing plant to pelletizing plant where the blend of additives is mixed with taconite concentrates and coal to be fired at 2400°F to produce a hard, dense pellet. Materials conveyed pneumatically are soda ash, lime, starch, bentonite, and pulverized coal.

Trash and Linen Collection

An application of pneumatic conveyors not usually thought of by engineers in industry is the conveying of trash and linens. Vacuum-type systems are used in hospitals to remove trash and soiled linens as soon as they are generated, providing sanitary conditions, eliminating unpleasant accumulations in closets or carts, and reducing labor and contamination. Standard systems are 16-in., but material conveying lines can be supplied in diameters up to 24-in. The systems are activated by a centrifugal turbocompressor with radial-bladed impeller and noise suppression damper. Separate systems for trash and linens are recommended for safety. It is claimed that individual units weighing up to 50 lb can be conveyed successfully in the 16-in. system.

Trash collection systems for large residential apartment complexes and multiple office buildings have been installed handling as much as 75 tons/day. Conveying pipes as large as 36 in. have been installed with the vacuum produced by centrifugal turbocompressors.

Long-Distance Pipelining

For the last 40 years much debate has highlighted the controversy between pneumatic and slurry long-distance pipelining. Regardless of the results of all the discussions, one characteristic of the material to be pipelined is of paramount importance: Is it water absorbant? If yes, slurry cannot be used; if no, slurry is acceptable; for pneumatic conveying, this characteristic is immaterial.

Of further importance is the power required and the length of the individual pipelines (no booster stations). It appears that slurry is more economical and capable of much longer pipelines than pneumatics. This author's recommendation is slurry yes, pneumatic no.

6 HANDLING CHARACTERISTICS OF MATERIALS

In conveying of materials by pneumatic conveyors, certain physical characteristics of the materials determine which type of system to use. The following descriptions give many of these characteristics as well as pertinent information as to why certain conveying systems can or cannot be applied in particular situations. Also, component selections are noted where their use is mandatory to successful operation. We will find that variances may be necessary because of system layout, conveying rates and timing, and type of operation.

Adipic Acid

Bulk density: 47 lb/cu ft. Particle size: fine and very fine. Conveyed in vacuum, low-pressure, and closed-circuit systems. Air-activated conveyor questionable, test for positive results. To avoid contamination, use stainless steel constructed equipment.

Alfalfa Meal

Bulk density: ground, 16 lb/cu ft; pelleted, 38 lb/cu ft. Particle size: ground, all passes a 14 mesh screen; pelleted, $\frac{1}{2}$ in. and under. Both types are conveyed by vacuum and low-pressure systems. The pelleted type requires extreme care to minimize pellet breakage and attrition. Design category, soft feeds.

Alum (Commercial Aluminum Sulfate)

Bulk density: powder, 42 lb/cu ft; ground, 63 lb/cu ft; rice, 60 lb/cu ft; lump, 65 lb/cu ft. Particle size: powder, all passes a 100 mesh screen; ground, $\frac{1}{8}$ in.

to 65 mesh screen; rice, $\frac{3}{16}$ in.; lump, $\frac{3}{4}$ in. and under. All types of alum are conveyed by vacuum and low-pressure systems and blow tanks. The powdered alum can be conveyed by an air-activated gravity conveyor. All types have a tendency to smear, so rotary feeders and steel-conveying pipeline bends should be avoided. Gate locks should be used below filter-receivers on vacuum systems and as line chargers for low-pressure systems. The pipeline bends should be reinforced rubber with a fabricated strong back to maintain an even curve.

Alumina

Bulk density: 50–60 lb/cu ft. Particle size: fine, with varying percentages passing through 200 and 325 mesh screens. Two types are commercially produced, floury and sandy. Although both types are easily fluidized for conveying by an air-activated gravity conveyor, each has its own characteristic regarding conveyability by various conveyors. Both, however, are chemically alike and extremely abrasive, about 8.8 on the Moh scale.

Floury alumina is somewhat sticky and has a high angle of repose. Undisturbed, it will stand up straight, but after disturbance and aeration it will stay soupy for quite some time. It is suitable for conveying by vacuum, low-pressure, the fluid solids pump, and blow tank.

Sandy alumina is free flowing and relatively easy to convey by vacuum from a railcar, barge, or ship. Unlike floury alumina, sandy alumina has a low angle of repose, and although it can be aerated, it loses its aerated state very rapidly. A minimum velocity for vacuum conveying is 115 ft/sec, whereas the floury type requires only 80–100 ft/sec. Conveying by vacuum is excellent, by low-pressure satisfactory, and by fluid solids pump difficult.

Equipment for handling alumina must be made to withstand abrasion. Rotary airlocks will wash out in a relatively short time. Gate locks or interlocked bottle containers should be used below filter-receivers. Rubber unloading hoses and wearbacks on all conveying pipeline bends are necessary.

In filter-receivers, when conveying sandy alumina, use a filter media of cotton construction, and when conveying the floury type, use nylon. Using either type of medium for both types of alumina will shorten its useful life.

Aluminum Hydrate

Bulk density: 55–75 lb/cu ft. Particle size: fine, with varying percentages passing through 200 and 325 mesh screens. Its conveyability parallels that of alumina. Vacuum systems have no problems, but when conveying through a dense stream using minimum velocities, care must be used in determining air requirements. A hydrate having 51.3% passing a 200 mesh screen and 37.4% passing a 325 mesh screen was found to require about a

30% higher velocity than a hydrate passing 96.6% through a 200 mesh screen and 79.1% through a 325 mesh screen. The coarser the material, the higher the velocity.

This material is abrasive, so equipment should be heavily constructed. Gate locks, fluid solids pumps, and blow tanks together with wear reinforcement of pipeline bends must be used.

Aluminum Oxide

Bulk density: 60–85 lb/cu ft. Particle size: fine and very fine. Extremely abrasive, between 8.0 and 9.0 on the Moh scale. This material is a prime candidate for the high-density pulse phase and low-velocity-type systems.

Ammonium Sulfate

Bulk density: 50–60 lb/cu ft. Particle size: granular, $\frac{3}{16}$ in. and under. The material is hygroscopic; maximum moisture content for conveying, 1%. Buildup will occur in feeders and cyclone receivers requiring cleaning at 3-month intervals. pH is extremely low, indicating high acid, requiring Orlon filter media or air washers for dust retention. Vacuum and low-pressure systems are used for conveying. Velocity must be strictly controlled to minimize attrition.

Arsenic

Bulk density: 125 lb/cu ft. Particle size: fine, slightly abrasive. Vacuum systems normally used to convey this material since dust losses are to be avoided. Closed-circuit systems ideally suited. In a vacuum system, as well as a closed-circuit, the velocity should be 125 ft/sec and the filter-receiver should be oversized. Some arsenic oxides can be conveyed by a fluid solids pump when their fineness is allowable.

Asbestos

Three forms of asbestos are conveyed by vacuum, low-pressure, and closed-circuit systems: shred, dust, and floats. Bulk density: 6–25 lb/cu ft. Particle size: irregular. All three types of asbestos can be classed as extremely light, causing difficulty in getting the material into or out of the airstream. The floats and dust are flakelike and mildly abrasive. Low velocities should be used in the conveying pipelines. Rotary feeders should have an extra large inlet and outlet to accelerate flow. In the inlet of the feeder, a baffle plate should be inserted to deflect the flow from the leading edges of the rotor blades, to avoid shearing, clipping, and possibly stalling caused by the shredded material getting caught between the rotor and body.

Barley

Bulk density: 38 lb/cu ft, 48 lb/bu. Conveyed in vacuum and low-pressure systems. Breakage to the kernel and skinning of the hulls are to be avoided by controlling pipeline velocity and constructing the pipeline with rectangular section bends having renewable wear backs and a slow-down section prior to entry into the terminal receiver.

Barytes (Barium Sulfate)

Bulk density: fine and very fine, 100–125 lb/cu ft; micronized (extremely fine), 65 lb/cu ft. To convey this material pneumatically, it must be fine, with a minimum of 75% passing through a 200 mesh screen and all through a 100 mesh screen, plus a good part of the amount passing through the 200 mesh screen, and also passing through a 325 mesh screen to add fluidity to the material. For continuous conveying, the fluid solids pump is most applicable. If intermittent conveying is permissible, medium- and high-pressure blow tanks will give a good performance. The air-activated gravity conveyor is applicable for horizontal conveying.

Bauxite

Bulk density: 60–70 lb/cu ft. Particle size: ground, dried, sufficiently fine for fluid solids pump and blow tank conveying. Extremely abrasive. Air-activated gravity conveyor can be applied provided the material is acceptable to this type of conveyor.

Beet Pulp, Dried

Bulk density: 16 lb/cu ft. Particle size: coarse to fine, fibrous and stringy. Conveyed by vacuum and low-pressure systems. For best results, use low vacuum as well as low pressure. Use baffle plate in throat of rotary feeder to avoid getting the material getting caught between the rotor and body.

Bentonite

Bulk density: 50–60 lb/cu ft. Particle size: 90% through a 200 mesh screen. Bentonite is a material similar in physical characteristics to clay or kaolin. It is slightly abrasive and hygroscopic. Being easily fluidized, it can be conveyed by any of the pneumatic-type systems. Air-activated gravity conveyors are excellent for horizontal conveying. What system to use will be determined by the fineness, moisture content, conveying rate, layout, and purpose of the system.

Bone Meal

Bulk density: 50 lb/cu ft. Particle size: 95% through a 10 mesh screen. When conveyed by either a vacuum or low-pressure system, this material should be steam-dried. Design category, soft feeds.

Borax (Sodium Tetraborate)

Bulk density: 50–65 lb/cu ft. Particle size: granular, from 20 to 140 mesh screens; dehydrated, fine. The material is nonabrasive and is conveyed by vacuum and low-pressure systems. In conveying through low-pressure systems, check the melting point. Some types of borax have a low melting point (in own water of crystallization) of 144°F. When conveying this type of borax, conveying air temperatures must be controlled.

Borax ore, known as rasorite or kernite, has a bulk density of 70 lb/cu ft and is granular and nonabrasive.

Anhydrous borax and rasorite have a bulk density of 72–89 lb/cu ft; they are granular, but very abrasive. For these materials equipment should be heavily constructed. Rotary airlocks should be avoided, gate locks should be used. Design category, salt cake.

Boric Acid

Bulk density: 52 lb/cu ft. Particle size: granular, from 20 to 200 mesh screens. This material is nonabrasive and mildly corrosive. For conveying, use vacuum and low-pressure systems. Maximum conveying air temperature, 200°F.

Bran

Bulk density: 16 lb/cu ft. Particle size: irregular; interlocks, difficult to pick up in boxcars for vacuum conveying, conveying rate will be approximately 40–50% of that attained on average soft feeds. Low-pressure conveying possible, but blowback through rotary feeder will affect conveying rate. For best results, use low vacuum as well as low pressure.

Brewers Dried Grains

Bulk density: 25–30 lb/cu ft. Particle size: all through 20 mesh screen. Some extra dry grains will have a bulk density as low as 15 lb/cu ft. Conveyed by vacuum and low-pressure systems. For best results, use low-pressure as well as vacuum systems.

Calcium Carbonate (Chalk, Limestone, Marble Dust, Whiting)

Calcium carbonate in its various forms has characteristics that affect its flow and conveyability, determining the system that can be applied.

Chalk, Whiting

Bulk density: 70–75 lb/cu ft. Particle size: all through 100 mesh screen. The material is mildly abrasive and becomes fluid under the influence of air. It can be conveyed by any of the pneumatic-type systems. Conveyance by an air-activated gravity conveyor is mostly good, except where an excessive amount of fines causes agglomeration.

Limestone

Bulk density: pulverized, 75 lb/cu ft; crushed, 85–90 lb/cu ft. Particle size: pulverized, all through a 100 mesh screen; crushed, $\frac{3}{8}$ in. and under. Both types are abrasive. Pulverized is conveyed by the fluid solids pump and blow tank, as well as the air-activated gravity conveyor. The crushed is conveyed by vacuum and low-pressure systems. All equipment must be heavily constructed. Rotary airlocks are to be avoided, gate locks must be used. When conveying pulverized limestone that is too coarse for a fluid solids pump, a high velocity and a light to medium saturation must be used to avoid settling of material out of the airstream in a low-pressure pipeline system. Pulverized can be handled in a high-density, pulse phase system.

Marble Dust

Bulk density: 80 lb/cu ft. Particle size: all through a 100 mesh screen. The material is abrasive and becomes fluid under the influence of air. The fluid solids pump and blow tank are applicable, as is the air-activated gravity conveyor.

Precipitated Calcium Carbonate

Bulk density: average 40–55 lb/cu ft; some grades 18–25 lb/cu ft. The material is nonabrasive, and although it tends to become fluid under the influence of air, it will agglomerate. Only vacuum and low-pressure systems are applicable for conveying.

Carbon (Activated)

Bulk density: 15 lb/cu ft. Particle size: all through a 325 mesh screen. Conveyed by vacuum and low-pressure systems and fluid solids pump. Low vacuums and pressures should be used as well as oversized filter-receivers.

Conveyance by air-activated gravity conveyors will depend on whether the carbon will remain flowable under the influence of air or will agglomerate.

Carbon (Carbon Black)

Bulk density: pelletized, 20–25 lb/cu ft; powder, 4–6 lb/cu ft. Particle size: pelletized, $\frac{1}{16}$ in.; powder, all through a 325 mesh screen. Neither of these materials should be conveyed by pneumatic conveyors. Pelletized carbon will be reduced to powder by attrition in the pipeline system, and powdered carbon will be most difficult to separate from the airstream in a filter-receiver.

Catalysts

Catalysts for petroleum refining vary in particle size from extra fine to spheres, $\frac{3}{4}$ in. in diameter. Bulk density will also vary to as high as 150 lb/cu ft. Some are extremely fragile, breaking on impact with a cement floor from a height of only 3 ft, as well as abrasive. Although all types of pneumatic conveyors are used, each catalyst must be judged by its ability to be conveyed, to withstand attrition, and to be conveyed by the system best suited to the application and material characteristics.

Cellulose Acetate

Bulk density: 15–22 lb/cu ft. Particle size: fine and flake. Heat sensitive. Conveyed by vacuum and low-pressure systems. Conveyance by air-activated gravity conveyor depends on particle size and shape.

Cement (Portland)

Bulk density: 70–94 lb/cu ft. Particle size: 95% through a 200 mesh screen. Abrasive; do not use rotary feeders or airlocks when a pressure differential exists between the inlet and outlet. The rotary feeder having high-hardness cast metal construction with wearing blocks in the upper quadrant may, however, give adequate service. Fluid solids pumps and blow tanks are recommended. Air-activated gravity conveyors are an excellent application for horizontal conveying. They were developed in and by the cement manufacturing industry.

Cerelose

Bulk density: 40 lb/cu ft. Particle size: coarse to fine. Conveyed by vacuum and low-pressure systems, blow tanks, and air-activated gravity conveyors. Back bevel rotor blades of rotary feeders.

Chalk

See Calcium Carbonate.

Citrus Pulp (Dried)

Bulk density: 20 lb/cu ft. Particle size: coarse, flaky. Conveyed by vacuum and low-pressure systems.

Clay (Kaolin)

Clay in various physical characteristics is used in the manufacture of ceramics, insecticides, paints, paper, plastics, and rubber. For each industry, its physical characteristics vary. Clays for insecticides, paints, paper, and plastics are not considered abrasive for pneumatic conveying. For ceramics, however, they are abrasive. The dividing line between being and not being abrasive is the amount of free silica. This will vary between the ball type and the hard and soft grades. Normally, any amount of free silica higher than 15% will cause the clay to be abrasive for pneumatic conveying.

When handling clay in the ceramics industry, the fluid solids pump and blow tank should invariably be used. For other industries, all types are applicable, with vacuum and low-pressure systems the most prevalent. Most clays can be conveyed on an air-activated gravity conveyor. Those whose fineness causes agglomeration and whose moisture content and particle shape resist aeration, will not flow.

Some clay in granular form is used as a filtration agent in the refining of oil. In conveying this type of clay pneumatically, some degradation and attrition may occur in the conveying pipeline. This may be detrimental to the filtration process. In cases such as this, what happens in the conveying system may affect the end use of the product. End use and its requirements should be investigated.

Bulk density: rubber grade air floated, 32–35 lb/cu ft; paper makers clay (see Table 5.2); ceramics and ball clays, 50–75 lb/cu ft; insecticide clay, 12–20 lb/cu ft; paint and plastics clays, 20–35 lb/cu ft. Particle size: granular to extra fine.

Coal

Bulk density: 50–60 lb/cu ft. Particle size: coarse to extra fine. Fluid solids pump, blow tank, and air-activated gravity conveyor excellent applications for conveying pulverized coal. Slack bituminous and sized anthracite and bituminous conveyed in low-pressure systems. All coal is mildly corrosive. Anthracite is mildly abrasive, bituminous nonabrasive, and run of mine at times very abrasive.

Cocoa

Cocoa beans should not be conveyed pneumatically. They are extremely fragile, even to the skins. When the skins are broken, rancidity occurs in storage. *Cocoa powder*, bulk density, 26 lb/cu ft, and extremely fine, can be conveyed by vacuum and low-pressure systems. It is, however, heat sensitive, so conveying air temperatures must be controlled. Filter media should have a ratio of no greater than 1:1.

Coffee

Bulk density: green bean, 32 lb/cu ft; roasted bean, 22–26 lb/cu ft; ground, 25 lb/cu ft; soluble, 19 lb/cu ft. Particle size: green and roasted bean, granular; ground, fine; soluble, very fine. All types conveyed by vacuum and low-pressure systems. Bean should be conveyed using minimum velocities to avoid breakage. Soluble is hygroscopic; systems may require periodic cleaning to avoid contamination.

Coke

Bulk density: 25–35 lb/cu ft. Particle size: fine and very fine. Cokes applicable to pneumatic conveying are screenings, coke fines or particle, and coke flour. All are highly abrasive. Vacuum and low-pressure systems are recommended. Blow tanks applicable. Fluid solids pump applicable for flour. Maintain minimum velocities and a dense conveying stream.

Coke (Petroleum)

Bulk density: 30–35 lb/cu ft. Particle size: fine. Highly abrasive. Vacuum, low-pressure, and blow-tank systems and air-activated gravity conveyors are applicable. Avoid rotary feeders, with the possible exception of the feeder of high-hardness cast metal construction with wearing blocks in the upper quadrant.

Copra

Bulk density: 22–25 lb/cu ft. Particle size: lumpy. Vacuum and low-pressure systems are applicable. Conveying rate in vacuum-type system unloading ships affected by interlocking nature of material.

Corn (Shelled)

Bulk density: 45 lb/cu ft, 56 lb/bu. Particle size: granular. Conveyed by vacuum and low-pressure systems. Attrition and breakage to be avoided by controlling pipeline velocity, use of rectangular section bends with renewable wearplate, and a slowdown section prior to entering the terminal receiver.

Corn Grits

Bulk density: 42 lb/cu ft. Particle size: fine. Conveyed by vacuum and low-pressure systems.

Cottonseeds

Bulk density: 22–40 lb/cu ft. Particle size: granular. Only delinted seeds applicable for vacuum and low-pressure systems.

Cottonseed Meal

Bulk density: 35–40 lb/cu ft. Particle size: fine. Conveyed by vacuum and low-pressure systems. Design category, soft feeds.

Diatomaceous Earth

Bulk density: 11–27 lb/cu ft. Particle size: extremely fine. Abrasive. Vacuum and low-pressure systems, the modified diaphragm pump, and the air-activated gravity conveyor applicable. For best results use low-operating vacuum as well as low-operating pressure. Size filter-receivers and dust collectors most generously.

Distillers Grains (Dried Solubles)

Bulk density: 25–30 lb/cu ft. Particle size: fine. Vacuum and low-pressure systems applicable.

Dolomite

Bulk density: 65–80 lb/cu ft. Particle size: coarse and fine. Abrasive in a pneumatic conveyor. Conveyed best in the pulverized state using fluid solids pump, blow tank, and air-activated gravity conveyor.

Feldspar

Bulk density: 60–70 lb/cu ft. Particle size: fine. Abrasive in a pneumatic conveyor. Vacuum type, fluid solids pump, blow tank, and air-activated gravity conveyor applicable. Avoid rotary feeders in vacuum-type system.

Ferric Sulfate

Bulk density: 50–75 lb/cu ft. Particle size: granular. Very hygroscopic. Suitable for vacuum conveying only.

Ferrous Sulfate

Bulk density: 50–75 lb/cu ft. Particle size: granular. Mildly abrasive, heat sensitive. Efflorescent in dry air, oxidizes in moist air. Suitable for vacuum conveying only.

Fish Meal

Bulk density: 30–35 lb/cu ft. Particle size: coarse to fine. Conveyed by vacuum and low-pressure systems. Design category, soft feeds.

Flaxseed

Bulk density: 45 lb/cu ft. Particle size: granular. Mildly abrasive. Hard-shelled. Conveyed by vacuum and low-pressure systems.

Flint

Bulk density: 50–65 lb/cu ft. Particle size: fine. Abrasive. Conveyed by vacuum system, fluid solids pump, blow tank, and air-activated gravity conveyor. Avoid rotary feeders in vacuum-type system.

Flour (Wheat)

Bulk density: 35–40 lb/cu ft. Particle size: extra fine. Conveyed by vacuum and low-pressure systems and air-activated gravity conveyor. No. 1 grade for baking purposes easily fluidized by air; lesser grades will result in gradual increase in sluggishness.

Fluorspar (Calcium Fluoride)

Bulk density: 80–90 lb/cu ft. Particle size: fine. Abrasive. Fluid solids pump, blow tank, and air-activated gravity conveyor recommended. High-velocity systems should be avoided.

Fly Ash

Bulk density: 35–45 lb/cu ft. Particle size: very fine. Abrasive. Vacuum system, fluid solids pump, blow tank, and air-activated gravity conveyor recommended.

Fuller's Earth

Bulk density: 35–40 lb/cu ft. Particle size: granular and fine. A clay used for filtration. *See* Clay (kaolin).

Gluten Meal

Bulk density: 30–38 lb/cu ft. Particle size: fine. Hygroscopic. Vacuum and low-pressure systems recommended. Design category, soft feeds.

Grits (Brewers)

Bulk density: 42 lb/cu ft. Particle size: coarse grits, $\frac{1}{8}$ in. and under; refined grits, extra fine. Vacuum and low-pressure systems recommended. Refined grits approach physical characteristics of pulverized starch.

Gypsum (Calcium Sulfate, Plaster of Paris)

Bulk density: 60–80 lb/cu ft. Particle size: coarse and very fine. Mildly abrasive. Vacuum and low-pressure systems recommended for coarse material. Fluid solids pump and blow tank for very fine material. Air-activated gravity conveyor applicable, provided material does not agglomerate. In pressure systems, moisture absorbed by the gypsum as it is conveyed through the pipeline system can influence physical and chemical change; use caution.

Hog Fuel (Bark)

Bulk density: varies according to moisture content. Particle size: 4 in. and under. For specific details, see Pulp and Paper Industry, Section 5.7.

Ilmenite Ore

Bulk density: 110 lb/cu ft. Particle size: very fine. Blow tank, fluid solids pump, and air-activated gravity conveyor recommended.

Kaolin

See Clay.

Lime (Burnt, Calcium Oxide)

Bulk density: lump and pebble, 56 lb/cu ft: hydrated, 22–40 lb/cu ft. Particle size: lump, 2 in. and under; pebble, $\frac{3}{4}$ in. and under; hydrated, very fine. Slightly abrasive in a pneumatic conveyor. Lump and pebble lime conveyed by vacuum and low-pressure systems and the blow tank. Hydrated lime can be conveyed by all types of systems, including the air-activated gravity conveyor.

Limestone

See Calcium Carbonate.

Linseed Oil Meal

Bulk density: 34 lb/cu ft. Particle size: fine. Conveyed by vacuum and low-pressure systems. Oil content may affect conveying rates. Design category, soft feeds.

Magnesium Oxide

Bulk density: 40 lb/cu ft. Particle size: fine and extra fine. Moisture and moisture content will affect conveyability. Use caution.

Malt

Bulk density: 34 lb/bu. Particle size: granular. Conveyed by vacuum and low-pressure systems. For specific details see Brewing and Distilling Industries, Section 5.3.

Marble Dust

See Calcium Carbonate.

Meat Scraps

Bulk density: 38 lb/cu ft. Particle size: coarse. Conveyed by vacuum and low-pressure systems. Pressure systems will require periodic cleaning to remove residue fat and tallow from the interior of the system. Steam cleaning suggested.

Milk (Powdered)

Bulk density: 20 lb/cu ft. Particle size: fine. Conveyed by vacuum and low-pressure systems. Bevel rotor blades of rotary feeders.

Nylon

Bulk density: 34 lb/cu ft. Particle size: Pelletized, coarse; powdered, fine. Conveyed by vacuum and low-pressure systems. Heat-sensitive.

Oats

Bulk density: 32 lb/bu. Particle size: granular. Conveyed by vacuum and low-pressure systems. Hulls abrasive. Construct pipeline with rectangular section bends having renewable wearplates and provide a slowdown section prior to entry into the terminal receiver.

Oat Flour

Bulk density: 20 lb/cu ft. Particle size: fine. Conveyed by vacuum and low-pressure systems. Extremely sluggish, even under the influence of aeration.

Ores (Pulverized)

Ores can be conveyed by the fluid solids pump, blow tank, and the air-activated gravity conveyor, depending on the bulk density and particle size and shape.

Peanuts

Bulk density: 17–24 lb/cu ft. Particle size: coarse. Fragile. Unshelled can be conveyed by vacuum and low-pressure systems. Low velocity must be closely controlled; avoid clipping in the rotary feeder, and provide slowdown section prior to entry into the terminal receiver. Although some have said they were successful, this author feels that shelled peanuts should not be conveyed pneumatically, unless you wish to have peanut butter discharge at the terminal end of the conveying system.

Perlite

Bulk density: 8–15 lb/cu ft. Particle size: fine bead. Fragile. Conveyed by vacuum and low-pressure systems. Velocity must be closely controlled. Use low operating vacuums and pressures. Exploded perlite, having a bulk density of 4–7 lb/cu ft, should be conveyed by a fan-propelled system.

Phosphate Rock (Ground)

Bulk density: 80–100 lb/cu ft. Particle size: fine. Slightly abrasive. Conveyed by fluid solids pump, blow tank, and air-activated gravity conveyor. Exercise care. Some rock not easily fluidized. Under the microscope, can be described as having irregular and interlocking surfaces.

Phosphates (Trisodium, Tetrasodium)

Bulk density: 50–65 lb/cu ft. Particle size: fine. Conveyed by all types of pneumatic systems. The air-activated gravity conveyor, depending on fineness.

Plaster

See Gypsum.

Polyethylene

Bulk density: 30–40 lb/cu ft. Particle size: pellets, $\frac{1}{4}$ in. and under; powder, fine. Conveyed by vacuum and low-pressure systems. Heat sensitive. For specific details, see Plastics Industry, Section 5.5.

Polyvinyl Chloride (PVC)

PVC by itself is a material having many characteristics. Its bulk density varies from 17 to 50 lb/cu ft; its particle size varies from all passing through a 325 mesh screen to medium-sized pellets. The pellets and material in the range of 48–150 mesh have excellent flow and conveyability characteristics. PVC is heat sensitive. Conveyed by vacuum and low-pressure systems. Extrafine PVC (all passing a 325 mesh screen down to 0.05 microns in diameter) tends to agglomerate under the influence of air, which makes flow and conveyability difficult. Fine PVC can be conveyed by the air-activated gravity conveyor.

   PVC is a thermally sensitive thermoplastic. To convert it to an end product, additives such as stabilizers, plasticizers, lubricants, impact modifiers, and fillers and pigments are required. When these are added, flowability and conveyability change. Corrections, either increased or decreased velocity and saturations, are then necessary.

   The following design figures are in the ball park. They should not be construed as collectively applicable. Each compound should be treated individually.

Velocity: Vacuum, 100 fps; Pressure, 55 fps.

Conveying Distance	Vacuum		Pressure		
	Sat.	hp/ton	Sat.	hp/ton	Factor
100 ft	2.2	3.3	1.5	2.3	3.0
150 ft	2.4	3.8	1.6	2.9	
250 ft	2.7	4.4	1.8	3.2	
400 ft	3.0	5.4	2.0	3.8	

For additional details, see Plastics Industry, Section 5.5.

Resin (Phenolic)

Bulk density: 15–20 lb/cu ft. Particle size: fine. Conveyed by vacuum and low-pressure systems. Heat sensitive. Explosive. Conveying air temperature should not exceed 100°F. For additional details, see Plastics Industry, Section 5.5.

Rice

Bulk density: 30–36 lb/cu ft. Particle size: coarse. Fragile. Conveyed by vacuum and low-pressure systems. Breakage of the raw and polished rice grain is serious to food processors. To convey pneumatically for minimum breakage, a low velocity must be strictly controlled and as dense a conveying stream as possible maintained. USDA defines a broken grain as a piece of the kernel that is less than ¾ of the length of a whole kernel, and a split kernel.

Rough rice with the hulls on (they contain about 20% opaline silica) is very abrasive. It will wear out pipelines and cyclone receivers in a relatively short time. Rotary feeders, however, will have a good life-span.

Rubber Pellets

Bulk density: 40 lb/cu ft. Particle size: 1 in. and under. Conveyed by vacuum and low-pressure systems. For additional details, see Rubber Industry, Section 5.6.

Rye

Bulk density: 56 lb/bu. Particle size: granular. Conveyed by vacuum and low-pressure systems. Attrition and breakage to be avoided by controlling pipeline velocity, use of rectangular section bends with renewable wearplate, and a slowdown section prior to entering the terminal receiver. Design category, wheat.

Salt (Sodium Chloride)

Bulk density: rock, 40–50 lb/cu ft; fine, 70–80 lb/cu ft. Particle size: rock, ½ in. and under; fine, all through 20 mesh screen. Slightly abrasive. Corrosive and hygroscopic. Conveyed by vacuum and low-pressure systems and blow tank. Only kiln-dried salt should be conveyed pneumatically. Solar-dried salt retains its water of crystallization, which is dissipated during conveying.

Salt Cake (Impure Sodium Sulfate)

Bulk density: 70–109 lb/cu ft. Particle size: coarse granular to extremely fine. Slightly abrasive. Hygroscopic but not deliquescent. Conveyed by vacuum and low-pressure systems. For additional details, see Pulp and Paper Industry, Section 5.7.

Sand

Bulk density: 90–100 lb/cu ft. Particle size: fine. Abrasive. Conveyed by vacuum and low-pressure systems and blow tank. All equipment must be heavily constructed. Rotary airlocks are to be avoided. Gate locks should be used in vacuum and low-pressure systems. Some sands applicable for conveying by the air-activated gravity conveyor.

Sawdust

Bulk density: 10–25 lb/cu ft. Particle size: coarse to fine. Conveyed by vacuum and low-pressure systems. On light sawdust, use low vacuums and pressures for conveying to restrict blowback through the rotary feeder. On combination vacuum-pressure systems, use two feeders between cyclone-receiver outlet of vacuum leg and intake tee of pressure leg. For additional details see Pulp and Paper Industry, Section 5.7.

Seeds

Conveyed by vacuum and low-pressure systems. Degradation and attrition must be minimized and closely controlled.

Semolina

Bulk density: 40–45 lb/cu ft. Particle size: fine. Conveyed by vacuum and pressure systems and air-activated gravity conveyor.

Silica (Pulverized, Flour)

Bulk density: 75 lb/cu ft. Particle size: very fine. Abrasive. Conveyed by fluid solids pump, blow tank and air-activated gravity conveyor.

Soaps and Detergents (Finished Products)

Subject to smearing and smudging when conveyed pneumatically. Suggest other types of conveyors. Satisfactory for air-activated gravity conveyor.

Soda, Anhydrous Caustic (Dry)

Bulk density: 60–70 lb/cu ft. Particle size: fine. Conveyed by low-pressure blow tank. Dry air required.

Soda Ash (Sodium Carbonate)

Synthetic ash (ammonia-soda process). Bulk density: light, 35 lb/cu ft; dense, 63 lb/cu ft. Particle size: light, very fine; dense, fine. High velocity in airstream required. If velocity is too low, the material will drag and build up, particularly in long horizontal pipeline runs and elbows (long radius bends). With high velocity, however, flow characteristics are altered and some degradation takes place. Pneumatic conveying is not recommended for dense soda ash where particle degradation cannot be tolerated.

Natural Ash (Green River, Wyoming)

Bulk density: 49 lb/cu ft. Particle size: 30–100 mesh screen, needlelike in structure. Not as fragile as synthetic ash. Vacuum and low-pressure systems recommended. Blow tanks applicable. Particle size and shape will determine conveyability by air-activated gravity conveyor.

Sodium Phosphates

Bulk density: 65 lb/cu ft. Particle size: fine. Conveyed by vacuum and low-pressure systems and blow tank.

Sodium Sulfate (Glaubers Salt)

Bulk density: 88 lb/cu ft. Particle size: fine. Conveyed by vacuum and low-pressure systems. Fineness will determine conveyability by air-activated gravity conveyor. Design category, salt cake.

Sodium Sulfite

Bulk density: 75–97 lb/cu ft. Particle size: fine. Hygroscopic. Erosive when entrained in air. Conveyed by vacuum and low-pressure systems. Design category, salt cake. For additional details see Pulp and Paper Industry, Section 5.7.

Soybeans

Bulk density: 46–48 lb/cu ft. Particle size: granular. Extremely abrasive in pneumatic conveying pipeline. Rotary feeders, however, will have a good life-span. Conveyed by vacuum and low-pressure systems.

Soybean Meal

Bulk density: 38 lb/cu ft. Particle size: fine. Solvent type easier to handle than expeller type because of lower oil content. Maximum oil content for good conveying rates, 13%. Conveyed by vacuum and low-pressure systems. Design category, soft feeds.

Starch (Corn)

Bulk density: pulverized, 40–50 lb/cu ft.; pelletized, 30–35 lb/cu ft. Particle size: pulverized, extra fine; pelletized, ¼ in. and under. Explosive. On car-unloading systems insert permanent magnet type magnetic separator to remove tramp iron. First pneumatic conveyor to prove its safety features in conveying starch, fluid solids pump. Conveyed by vacuum and low-pressure systems, fluid solids pump, and air-activated gravity conveyor. Back bevel rotor blades of rotary feeders.

Starch (Potato)

Bulk density: 40–45 lb/cu ft. Particle size: extra fine. A very difficult material for pneumatic conveying. In vacuum conveyors, feeder clearances had to be enlarged considerably to avoid binding. In low-pressure systems, material buildup occurs in rotary feeders and conveying pipeline. Friction between the material and equipment causes gelatinization.

Sugar (Granulated)

Bulk density: 50–55 lb/cu ft. Particle size: fine. Moisture sensitive. Attrition in conveying pipeline and rotary feeders may affect final usage. Back bevel rotor blades of rotary feeders. Sugar dust explosive. Conveyed by vacuum and low-pressure systems, blow tank, and air-activated gravity conveyor.

Raw sugar, because of its moisture content and stickiness, cannot be conveyed pneumatically.

Talc (Hydrous Magnesium Silicate, Soapstone)

Bulk density: 46–62 lb/cu ft. Particle size: extra fine. Conveyed by vacuum and low-pressure systems, fluid solids pump, and air-activated gravity conveyor. May build up on interior of conveying pipeline.

Titanium Dioxide

Bulk density: 45–55 lb/cu ft. Particle size: extra fine. Extremely abrasive. One of the most controversial materials for the application of pneumatic conveyors. Flowability, unpredictable. Both fluidization and momentary vibration required to induce flow. Continuous vibration compacts the material, making flow more difficult. Tends to build up on interior of conveying pipeline. Blow tank as applied to the pressure-differential self-unloading railroad car and truck, and low-pressure system has been successfully used. Rubber hose and aluminum and stainless steel pipe in use as conveying pipeline.

Wheat

Bulk density: 48 lb/cu ft, 60 lb/bu. Particle size: granular. Conveyed by vacuum and low-pressure systems. Attrition and breakage to be avoided by controlling pipeline velocity, use rectangular section bends with renewable wearplates, and a slowdown section prior to entering the terminal receiver.

Whiting

See Calcium Carbonate.

Wood Chips

Bulk density: 15–25 lb/cu ft. Particle size: very coarse, will interlock. Conveyed by low-pressure systems. For additional details, see Pulp and Paper Industry, Section 5.7.

Wood Flour

Bulk density: 10–15 lb/cu ft. Particle size: fine. Explosive. Conveyed by vacuum and occasionally, low-pressure systems. For additional details, see Plastics Industry, Section 5.5.

Zinc Oxide

Bulk density: 30–45 lb/cu ft. Particle size: fine. Can be sluggish at times. Conveyed by vacuum and low-pressure systems, fluid solids pump and blow tank.

BIBLIOGRAPHY

GENERAL

Ball, D. G., "Pneumatic Backfilling," ASME Paper No. 69-MH-22, 1969.

Billig, Joseph A., "The Fuller-Kinyon System: Solution to 1920s Problem Is at Center of New Fuel Technology," *Journal of Commerce*, November 16, 1981.

Burgess, Gordon, "Pulse Phase Conveying—A Review of the First Ten Years," *Bulk Solids Handling*, Vol. 1, No. 1, February 1981.

Dawson, James, "How Pneumatic Conveying Aids Safety," *Material Handling Engineering*, December 1961.

Emery, R. B., "Aeration Apparatus Converts Bin from Funnel Flow to Mass-Flow Characteristics," ASME Paper No. 72-MH-5, 1972.

Gerchow, Frank J., "How to Select a Pneumatic Conveying System," *Chemical Engineering*, February 17, March 31, 1975.

———, "Continuous Vacuum Pressure Systems," ASME Paper No. 80-WA/MH-2, 1980.

———, "Rotary Valve Selection for Pneumatic Conveying Systems," *Bulk Solids Handling*, Vol. 1, No. 1, February 1981.

Jolley, F. Raby and H. Walder, "Pneumatic Handling," *Mechanical Handling*, March and April 1949.

Knill, Bernie, et al., "Flow Lines—the Bulk Handling Game," *Material Handling Engineering*, September 1966.

Leitzel, Russel E., "Air-Float Conveyors," Pneumotransport 4, June 26–28, 1978, BHRA Fluid Engineering, Cranford, Bedford, England, Vol. 1, pp. D3 41–50, 1978.

Massey, H. C., "Pneumatic Handling Techniques for Bulk Commodities," presented at the ICHCA Bulk Transport Conference, Rotterdam, The Netherlands, June 1977.

Murphy, James D., "Materials to Resist the Abrasion of Pneumatically Transported Processed Refuse," *Energy Conservation Through Waste Utilization*, ASME, New York, 1978.

Overman, J. P. and T. G. Statt, "Evaluation of the Performance for a Pneumatic Trash Collection System," ASME Paper No. 77-RC-1, 1977.

Sayre, Howard S., "Selection and Use of Pneumatic Conveyors," *Material Handling Engineering*, September 1959.

Short, James C., "Modulode: Sensitive System for Pressure Conveying," *Rock Products*, April 1970.

Smith, D., "Pneumatic Transport and Its Hazards," *Chemical Engineering Progress*, Vol. 66, No. 9, September 1970.

Solt, Paul E., "Troubleshooting Pneumatic Conveying Systems," *Chemical Engineering Progress*, March 1976.

Stankovich, I. D., "A Systems Approach to Pneumatic Handling," in Frank J. Loeffler, Ed., *Unit and Bulk Materials Handling*, ASME, New York, 1980, pp. 163–169.

Stoess, H. A., Jr., "Match Material to System for Successful Pneumatic Conveying," *Material Handling Engineering*, April 1961.

———, "Pneumatic Pipelines," *Journal of Pipelines*, Vol. 1, No. 1, pp. 3–10, January 1981.

Westerfelt, Richard P., "Planning and Maintaining Pneumatic Conveying Systems," *Plant Engineering*, November 1966.

Zilly, Robert G., "Air Conveying," *Modern Materials Handling*, December 1957.

DESIGN

Crawley, M. F., "Practical Aspects of Modern Pneumatic Conveying Systems," ASME Paper No. 79-WA/MH-10, 1979.

Fenton, Donald L., "Pneumatic Conveying Overview," Proceedings of the Technical Program, International Powder and Bulk Solids Handling Convocation, Philadelphia, May 15–17, 1979, Industrial & Scientific Conference Management, Inc. Chicago.

Fischer, John, "Practical Pneumatic Conveyor Design," *Chemical Engineering*, June 2, 1958.

Hudson, Wilbur G., *Conveyors and Related Equipment*, 3rd ed., Wiley, New York, 1954, pp. 157–190.

Lempp, Matthias, "Physical Fundamentals for the Calculation of Pneumatic Conveyor Systems," *Deutsche Hebe und Foerdertechnik in Dienste der Transportationalisierung*, January 1962 (9), pp. 21–26.

Lyons, A. L., "Analyzing Pneumatic Conveying Systems," *Automation*, Vol. 18, No. 1, January 1971.

Leung, L. S. and Robert J. Wiles, "A Quantitative Design Procedure for Vertical Pneumatic Conveying Systems," *Ind. Eng. Chem.*, *Process Design and Development*, Vol. 15, No. 4, 1976.

Perkins, D. E. and J. E. Wood, "The Design and Selection of Pneumatic Conveying Systems," ASME Paper No. 75-MH-1, 1975.

Soo, S. L., "Design of Pneumatic Conveying Systems," *Journal of Powder & Bulk Solids Technology*, May 1980.

Weber, M., "Principles of Hydraulic and Pneumatic Conveying in Pipes," *Bulk Solids Handling*, Vol. 1, No. 1, February 1981.

Young, Henry T., "Design and Application of Closed Loop Pneumatic Conveying Systems," ASME Paper No. 61-BSH-6, 1961.

APPLICATIONS

Anon, "Floating Pneumatic Grain Elevators," *The Dock & Harbour Authority*, August 1953.

———, "Pneumatic Unloading Plant at Kittimat," *The Dock & Harbour Authority*, June 1955.

———, "Pneumatic Conveying Solves Taconite Processing Problems," *Engineering & Mining Journal*, Vol. 160, No. 10, October 1959.

———, "World's Longest Pneumatic Malt Handling System Installed by Dow in Montreal," *Modern Brewery Age*, December 1959.

———, "Burgermeister Expands UP with Pneumatic Handling," *Western Material Handling*, October 1960.

———, "Unique Handling Techniques Used to Automate New Georgia Clay Processing Plant," *Pit and Quarry*, September 1961.

———, "Switch from Bin to Bulk Handling More Than Triples Storage Capacity for Cranberries," *Food Processing*, February 1969.

———, "Pneumatic Transportation of Sand at Thomas Saeger," *Foundry Trade Journal*, September 13, 1973.

———, "Materials Handling," A Modern Plastics Special Report, *Modern Plastics*, May 1982.

Blanchard, Charles T., "Pressured Air Simplifies Conveying Sewage Solids," *Water and Sewage Works*, November 1976.

Butters, G., *Plastics Pneumatic Conveying and Bulk Storage*, Applied Science Publishers, London, 1981.

Carlson, Robert D., "Bulk Handling of TiO_2—Here's How It's Done," *Process Engineering News*, October 1976.

Dallaire, Gene, "Pneumatic Waste Collection on the Rise," *Civil Engineering—ASCE*, August 1974.

Drever, J. Bruce, "Bulk Flour Deliveries," *Baking Industry* 1954.

Fischer, John, "Pneumatic Conveying of Granular Plastics," *Chemical Engineering Progress*, January 1962.

Geisenheyner, Robert M., "Dry Bulk Trailer Equipment," ASME Paper No. 61-BSH-10, 1961.

Hall, Donald D., "Automated Broke Handling Can Improve Finishing Room Operation," *Pulp & Paper*, November 1981.

Heintjes, Harold, "Pneumatic Feeding Systems," *Coal Process Technology*, New York: AIChE, Vol. 5, pp. 236–239, 1979.

Jones, Don C., "Pneumatic Handling of Magnetite," *Mechanization*, August 1960.

Karstens, Jerry, "Production Costs Can Be Reduced," *Feed Age*, February 1955.

Poesch, Heinz, "Pneumatic Unloading Equipment for Bulk Cement Carriers," *Bulk Solids Handling*, Vol. 2, No. 2, June 1982.

Roberson, James E., "Bark Burning Methods," *Tappi*, Vol. 51, No. 6, June 1968.

Singh, Tara, "Pneumatic Waste Collection," *Pollution Engineering*, Vol. 6, No. 3, March 1974.

Soo, S. L., J. A. Ferguson and S. C. Pan, "Feasibility of Pneumatic Pipeline Transport of Coal," ASME Paper No. 75-ICT-22, 1975.

Stancato, A. J., "Powders Mean Profit, But They Must Be Tamed," *Modern Plastics*, November 1981.

Stoess, H. A., Jr., "Properties, Storage and Conveying of Salt Cake," *Tappi*, Vol. 41, No. 4, April 1958.

———, "Conveying and Storage of Clay," *Tappi*, Vol. 44, No. 6, June 1961.

Teichmann, Arno, "Pneumatic Ship Unloading Plants for Poorly Flowing Bulk Materials," *Bulk Solids Handling*, Vol. 2, No. 2, June 1982.

Thirsk, A. C. S., "The Bulk Handling of Flour and Wheatfeed," N.J.I.C. Gold Medal Thesis, Part V(a). Design and Layout of Bulk Storage Installations, *Milling*, September 7, 1962.

Trauffer, Walter E., "Texas Industries' Cement Plant in Dallas-Fort Worth Area," *Pit & Quarry*, March 1962.

Weller, L. G., "Flour and Sugar Pneumatically Handled at World's Largest Bread Bakery," *Bakers Weekly*, May 18, 1953.

Yates, J. P., "Bulk Cement Handling at Boulder Dam," *Western Construction News*, August 1934.

INDEX